ARTIFICIAL INTELLIGENCE THROUGH SIMULATED EVOLUTION

ARTIFICIAL INTELLIGENCE THROUGH SIMULATED EVOLUTION

Lawrence J. Fogel, Ph. D.
President, Decision Science, Inc.

Alvin J. Owens, Ph. D.
Vice President, Decision Science, Inc.

Michael J. Walsh, Ph. D.
Secretary-Treasurer, Decision Science, Inc.

JOHN WILEY & SONS, INC., NEW YORK • LONDON • SYDNEY

Library of Congress Catalog Card Number: 66-21067
Printed in the United States of America

To Our Wives

Preface

Life on earth has evolved for more than two billion years. Initially, only the strongest creatures survived the adversity of the natural environment and the competition of other living organisms. As time passed, certain species grew sufficiently complex to recall the repetition of similar events. As a result, these creatures developed skills, and soon the rule became "survival of the most skillful." That individual survived which could consistently sense danger, evade threat, or utilize camouflage. Indeed, the existence of man stands as testimony to the ability of his forebears to outwit, rather than out-power, in the competition for survivial. Clearly, the ability to survive was measured in terms of how consistently an individual made the proper choice from among the set of available alternatives.

With increased sensory capability, it became possible for creatures to recognize the existence of new situations that required decisions outside the realm of "conditioned reflex." The individual had to draw upon whatever analogies could be found between the observed situation and those in which his repertoire of available skills had previously demonstrated its worth. This was the beginning of what might be called intelligent behavior. Under the rigors of continual struggle, it was often the more intelligent creature that survived to reproduce more of its own kind. Some species "solved" the problem of survival by having an increased number of offspring, whereas others decreased the scope of their activity in order to exclude the apparent threat. Although these are of interest, particular attention will be devoted to "creatures" whose response to the threat of their environment is through intellectual adaptation.

This ability to discover aspects of similarity also provides a capability for self-observation and self-appraisal. Interest in making wise decisions was a natural outgrowth. Only in modern times has this interest been accompanied by an inquiry into the very nature of the decision-making process itself.

In recent years, the growth of technology has provided mankind with new tools, tools that may make it possible for him to synthesize automata which display those same features characteristic of his own decision-making behavior. The modeling of man may result in mechanisms that replicate his capability but perform with much greater speed, accuracy, and consistency.

But it is of far greater interest to search for an understanding of the logical properties that constitute intellect. One means toward this end is the study of the natural process that has evolved creatures of greater and greater intelligence. Natural evolution is recognized to be an interative process consisting of mutative reproduction, natural selection, voluntary recombination, individual learning, and so forth. It is hoped that through a replication of specific aspects of evolution, means will be found for the generation of artificially intelligent automata; automata capable of solving problems in previously undiscovered ways. This may well provide a deeper understanding of the very organization of intellect.

This book could not have been written without the encouragement of a number of individuals. In particular, we should like to acknowledge the interest in this research demonstrated by Dr. H. F. Dunholter, and the helpful technical comments offered by Drs. J. Lyman, J. Barnes, W. Adey, and C. Bell. The research was supported in part by General Dynamics, The Office of Naval Research of the USN, Goddard Spaceflight Center of NASA, and the Rome Air Development Center of the USAF. Special appreciation is due to Miss Mary Ann Downey for her cooperation in the preparation of the manuscript.

LAWRENCE J. FOGEL
ALVIN J. OWENS
MICHAEL J. WALSH

San Diego, California
April, 1966

Contents

Prologue

This book is intended to provide an overview of current research on the use of evolutionary programming in creating artificial intelligence. The problem of defining intelligence is considered (in Chapter 1) from the bionic and heuristic programming points of view. Fast-time simulation of evolution is offered as an alternative means for generating intelligent behavior. Chapter 2 describes the principles of evolutionary programming in some detail. For the sake of demonstration, intelligence is viewed as a composite of an ability to predict the environment, coupled with an ability to select a suitable response from the set of alternatives in the light of the prediction and the goal. Chapter 3 presents the results of a series of prediction experiments conducted in 2-symbol and 8-symbol languages.

Chapter 4 discusses the use of evolutionary programming as a means of attacking some of the fundamental problems of information: detection, discrimination, pattern recognition, and pattern classification. Special attention is devoted to the problem of interdiction—the diagnosis of a concurrent variable on the basis of the available evidence and history. Chapter 5 considers the problem of how evolutionary programming can be used as a means for controlling a plant even in the face of incomplete knowledge and uncertainty regarding its performance. Chapter 6 presents a discussion of some important implications of evolutionary programming. The realization of self-programming machines and the automatic conduct of the scientific method are considered. Some of the philosophic implications of the research are also mentioned, including the mechanization of logical induction, the creation of higher levels of abstraction, and a mechanization of self-awareness.

Chapter 7 offers a brief summary of the state of the art, with emphasis on the limitations of evolutionary programming. The Epilogue calls attention to certain important social problems that are created through the automation of intellect.

Throughout the book, specific references are carried as footnotes to the text. A list of general references is provided in the Bibliography. An appendix contains a logical description of an evolutionary program suitable for use on conventional digital computer equipments.

The purpose of the book will have been realized if this research is brought to the attention of the scientific community, in the hope of encouraging an extension of the results already attained; to the engineering audience, in the hope of finding a wide range of applications for this new computer capability; and to the general public, in the hope of directing its attention to problems that will require knowledgeable consideration in the near future.

Lawrence J. Fogel
Alvin J. Owens
Michael J. Walsh

ARTIFICIAL INTELLIGENCE THROUGH SIMULATED EVOLUTION

CHAPTER ONE

Introduction

DEFINITION OF INTELLIGENCE

Interest in intelligent decision-making has existed for a very long time. During this time, effort was first devoted to aiding the decision-maker, furnishing him with sound advice, and providing him with relevant information. Decision-making was treated as an art. Not until the latter part of the last century was any serious inquiry made into the behavioral aspects of intelligent decision-making. In the words of W. S. Jevons (1873), "Economy of mental power may be considered one of the main conditions on which our elevated intellectual position depends."[1] The attribute called "intelligence" was of concern, but it remained without precise definition.

At about this time, a new emphasis was placed upon the value of quantitative description in science. As Lord Kelvin (W. Thomson) remarked,

"I often say that when you can measure what you are speaking about, and express it in numbers, you know something about it; but when you cannot measure it, when you cannot express it in numbers, your knowledge is of a meagre and unsatisfactory kind; it may be the beginning of knowledge, but you have scarcely, in your thoughts, advanced to the stage of science, whatever the matter may be."

The distinction between knowledge and intelligence became clear; knowledge being the useful information stored within the individual, and intelligence being the ability of the individual to utilize this stored information in some worthwhile (goal-directed) manner.

[1] From *The Principles of Science; A Treatise on Logic and Scientific Method* (originally published 1874). Reprint, Dover Publications, New York, 1958.

Early attempts to measure intelligence grew out of a practical need to recognize those who deserved special education. Various intelligence tests were developed, and although the testing technique was successively improved and refined in the decades that followed, the interpretation of the resulting score as a measure of intelligence still remains open to serious question. For example, it is generally recognized that the widely accepted I. Q. tests currently in use favor those individuals who come from an urban culture and those who are more prone to verbal expression. Such relative measures of intelligence do not pretend to offer insight into the fundamental nature of intellect.

In recent years, there have been a number of attempts to gain a deeper understanding into the nature of intellect. Biophysical studies have provided gross identification of the function of sections of the cortex, yet little is known concerning the nonspecific brain mechanisms.[2, 3] There remains a wide gap between the current state of the art and a detailed understanding of how the physical embodiment provides intellect.

On the psychological side, efforts have been made to factor intelligent behavior into specific "components" that facilitate identification and individual measurement. Although such factors as *verbal comprehension, fluency, perception, psychomotor coordination, number, memory,* and *reasoning*[4] have been described in some detail, they are certainly not independent aspects of intelligent behavior. Intelligent behavior is also exhibited by creatures at much lower levels in the phylogenetic series. Are the foregoing factors also components of the behavior of these creatures? Such an approach to the definition of intelligence appears fraught with danger.

Such descriptions of intelligence are overly restrictive. Might it not be more meaningful to define intelligence in terms of the behavior of any goal-seeking entity, to measure the degree of intelligence in terms of the adequacy of its decision-making? Certainly, without the existence of a goal, decision-making is pointless, and the term "intelligence" has no meaning. More carefully stated, intelligence can be viewed as *the ability of any decision-making entity to achieve a degree of success in seeking a wide variety of goals under a wide range of environments.* The difficulty in carrying this definition into quantitative terms rests with the problem of defining what is meant

[2] *Biological and Biochemical Bases of Behavior,* edited by H. F. Harlow and C. N. Woolsey, The University of Wisconsin Press, Madison, 1958.

[3] D. E. Wooldridge, *The Machinery of the Brain,* McGraw-Hill, New York, 1963.

[4] B. F. Green, Jr., "Intelligence in Computer Simulation," *Transactions of the New York Academy of Sciences,* Serial II, Vol. 27, November, 1964, pp. 55–63.

by "degree of success," "variety of goals," and "range of environments." The decision-maker who can achieve a particular goal only in a certain environment and not in another displays little intelligence. *Versatility* is the essential ingredient of intelligent behavior. Under this definition, only a comparative evaluation of intelligence can be meaningful.

AVENUES TOWARD ARTIFICIAL INTELLIGENCE

Interest in decision-making has steadily widened. As effort to measure the intelligence of the decision-maker has progressed, there has been a concurrent inquiry into the logical structure of intellect. This inquiry has grown out of purely philosophical speculation into the mechanization of various hypotheses with a view to demonstrating their validity. It is now generally agreed that a deeper understanding of the organization of intellect can be demonstrated by constructing a device that exhibits what is recognized to be "intelligent behavior." The last decade has seen a significant surge of interest in such mechanizations.

These efforts may be loosely classified into two groups. The first group represents investigators who recognize the human animal to be the most intelligent creature in nature. It would appear worthwhile, therefore, to replicate human intellect in some specific manner. From a biological point of view, an attempt is made to synthesize a "neural network" that exhibits properties of the nervous system. The term "bionics" is generally applied to this work.

Much of the work in bionics is based on the fundamental insight, contributed by McCulloch and Pitts in 1943,[5] that any "functioning" which can be defined logically, strictly, and unambiguously in a finite number of words can also be realized by a network of "formal neurons." Another classic paper[6] (1959) provided a demonstration of the selective properties of the retina of the frog. Extending this work, von Foerster was responsible for a clear explication of the possible property filters comprised of linearly interconnected formal neurons.[7] Such research is being carried on to gain greater understanding of

[5] W. S. McCulloch and W. H. Pitts, "A Logical Calculus of the Ideas Immanent in Nervous Activity," *Bulletin of Mathematics and Biophysics,* Vol. 5, 1943, p. 115.

[6] J. Y. Lettvin, H. R. Maturana, W. S. McCulloch, and W. H. Pitts, "What the Frog's Eye Tells the Frog's Brain," *Proceedings of the IRE,* Vol. 47, November, 1959, pp. 1940–1951.

[7] H. von Foerster, "Bio-Logic," in *Biological Prototypes and Synthetic Systems,* Vol. 1, edited by E. E. Bernard and M. R. Kare, Plenum Press, New York, 1962, pp. 1–13.

data reduction and encoding at higher and higher levels in the phylogenetic series. The problem becomes significantly more difficult as the research concerns man.

Quite independently, Hebb suggested a brain model in which association cells were connected together in a complex, randomly organized network.[8] These connections existing prior to each successful decision were presumed to be made more permanent, thereby providing "memory traces." Farley and Clark simulated such a system[9] using

". . . a randomly interconnected network of nonlinear elements, each element having a threshold for incoming excitation, below which no action occurs, and above which the element 'fires.' When an element fires, its threshold immediately rises to infinity, and then, after a short refractory period, falls exponentially back toward its quiescent value. Furthermore, a short time after firing, an element transmits excitation to all other elements to which it is connected. The effectiveness of the excitation thus transmitted to a succeeding element is determined by a property of the particular connection known as its 'weight.' In general, there will be several incoming connections at any element, each having its individual weight . . . . At the instant of transmission the appropriate weight is added to any excitation already present at the succeeding cell. Thereafter the excitation decays exponentially to zero. Thus the model contained both 'spatial' and 'temporal' summation. If at any time this excitation exceeds the threshold of the succeeding element, that element performs its own firing cycle and transmits its own excitations . . . . Thus the system organized itself to distinguish between two distinct input patterns."[10]

Rochester used even simpler "neurons" to obtain similar results, while attempting to realize Hebb's model more closely.[11]

In still later work, Rosenblatt considered a random network organized in a somewhat different manner.[12] His Perception consisted of a set of sensory elements (*S*-units), which are normally connected to

[8] D. O. Hebb, *The Organization of Behavior,* John Wiley and Sons, New York, 1949.

[9] B. G. Farley and W. A. Clark, "Simulation of Self-Organizing Systems by Digital Computer," *Transactions of the IRE,* Vol. PGIT-4, 1954, pp. 76–84.

[10] G. G. Farley, "Self-Organizing Models for Learned Perception," in *"Self-Organizing Systems,"* edited by M. C. Yovits and S. Cameron, Pergamon Press, New York, 1960, pp. 7–31.

[11] N. Rochester, J. H. Holland, L. H. Harbt, and W. L. Duda, "Tests on a Cell Assembly Theory of the Action of the Brain, Using a Large Digital Computer," *Transcription of the IRE,* Vol. PGIT-2, 1956, p. 80.

[12] F. Rosenblatt, "The Perceptron: A Probabalistic Model for Information Storage and Organization in the Brain," *Psychological Review,* Vol. 65, 1958, p. 386.

a set of multilevel memory elements called association elements (A-units). The outputs of the A-unit are in turn randomly connected to a group of response elements (R-units), which read out the results of the pattern-recognizing operation. These R-units are connected via inhibiting lines to other R-units, and, by feedback paths, to the A-units. The experiments consist of two phases: a learning period and a test period. During the learning period, a number of stimuli are offered to the sensor, and each in turn is associated with one of the responses by forcing the appropriate subset to become dominant; that is, by artifically activating the appropriate response signal. Upon completion of the learning period, the Perceptron is shown a test stimulus and allowed to react without any control over the response. If the connectivity within the network has been sufficiently modified, the response will have a greater probability of being correct. In essence, a set of multilevel association elements is connected to the sensor and response cells. The excitation of the sensory elements by new stimulus patterns causes a particular distribution of activity in the association elements, which modifies their transfer property in some way. If, following this modification, a response element operates either spontaneously or by being forced, then a new distributed connectivity is created; so that, whenever the original stimulus pattern is again presented, it will correlate with this distribution in such a way as to favor the operation of the previous associated response.

To introduce learning into the model, it is necessary to go further and postulate that the distribution of ambient connectivity values over the network at any time represents the total experience of the system up to that time. If a system is to create a particular internal distribution in response to a combination of external stimuli and its own responses to these stimuli, it is necessary to postulate a process whereby information regarding the system's responses can be fed back to the system, this being a positive or negative reinforcement.

A fundamental limitation exists with any such device. A learned configuration of the logical network can be used to translate any symbol of the input alphabet into any symbol of the output alphabet; but as long as the learned configuration is maintained, this translation is invariant. In other words, conditional dependencies which lie beyond the alphabet of the input symbols cannot be recognized.

Of course, it is possible to construct multilayer systems that promise greater learning capability.[13] All of these, however, require instruction as to the desired stimulus-response pairs to be learned by the machine.

[13] F. Rosenblatt, *Principles of Neurodynamics,* Spartan Books, Washington, D. C., 1962.

Other techniques have been developed in which the outputs of threshold devices are combined to provide a weighted judgment of the classification of the stimulus pattern; however, in the interest of brevity we will not attempt a detailed description of them.[14, 15]

Such biological modeling can indeed serve a useful purpose. It can provide greater insight into the functioning of the biological mechanisms by directing attention to new aspects that deserve observation and study. Although a distinct measure of success has already been attained in this regard, it is difficult to find biological mechanisms that serve as "natural prototypes" for the design of new engineering products, least of all machines that exhibit artificial intelligence.

This is not surprising. The hope of realizing artificially intelligent automata through direct biological modeling would appear to be particularly unpromising because of the sheer complexity of biological information-processing systems. The central nervous system of man is an immensely complex structure, containing some 15 billion neurons. Of these, about 10 billion are included in the brain proper. Each neuron has a great number of dendrites able to furnish it with information. Even gross patterns of connection in the central nervous system are difficult to trace. The detailed organization of the nervous system is far from what might be called "readily available knowledge." Until recently, the neuron was viewed as a two-state device; however, it is now widely recognized that there are graded responses at each synapse. A great deal remains unknown.

In addition to a lack of information regarding the structure of the nervous system, there are significant difficulties associated with describing its functioning. For example, neural encoding remains far from decipherable in terms of specific representations. The information storage mechanism is of great interest, yet there remains a considerable difference of opinion as to its biological embodiment. Here, for example, it might well be that an understanding of the mechanism of memory requires not only specific information regarding neuronal activity, but also the actions of the proximal glial cells, the modifications imposed upon the affected molecular structure of protein molecules, and, possibly, other aspects.

Biologists are the first to admit a serious lack of knowledge in these areas. Attempts to model specific aspects of biological information

[14] G. S. Sebestyen, *Decision-Making Processes in Pattern Recognition,* The Macmillan Co., New York, 1962.

[15] B. Widrow, "Generalization and Information Storage in Networks of Adeline 'Neurons'" in *Self-Organizing Systems—1962,* edited by M. C. Yovtiz, G. T. Jacobi, and G. D. Goldstein, Spartan Books, Washington, D. C., 1962, pp. 435–462.

processing are certainly worthwhile as a step toward gaining additional knowledge of the mechanism. But at the same time, such models appear to be particularly unlikely to produce any artificially intelligent automata, at least within the foreseeable future.

The second group of investigators agree that man, because of his intelligence, is most worthy of modeling; however, this group prefers to view man as a psychological entity.[16] Such modeling may be accomplished in the following manner: A game situation is presented to the subject who is required to respond with a "move." His decision as to which move to make is then analyzed in order to reveal some consistent set of subquestions which facilitate his decision-making. Consistent aspects of his rationale are thus identified and these become the basis for a computer program that results in a machine capable of playing the game in a similar manner. This approach to artificial intelligence, generally called "heuristic programming," attempts to find a set of information reduction rules which, when applied to the observed data, will, in a relatively efficient manner, successfully transform these data into a desirable response. Through such analyses, the investigator hopes to find not only means for efficient data reduction through replication of the human decision-making process, but also certain features common to all these programs which might enable the design of a decision-maker which would prove to be "intelligent" even in the face of new problems and unexpected situations.

From a less specific approach, the flexible behavior of an intelligent decision-maker may be modeled in terms of adaptive servo systems[17] and sequential statistical decision procedures.[18, 19]

In the case of the adaptive servo systems, the intent is to provide a hierarchy of feedback systems so that the parameters of lower-ordered systems will be suitably modified as a result of the response behavior of the higher-ordered systems. The domain of "adaptability" is constrained by the configuration of the multiloop servo system and

[16] In many situations, replication of specific details may prove to be less effective than the modeling of overall logical properties; note the fact that although modern aircraft obey the same aerodynamic laws as do birds, they are not ornithopters and do not use feathers.

[17] W. R. Ashby, *Design for a Brain,* John Wiley and Sons, New York, 1952 (rev. ed. 1960).

[18] K. S. Fu, "A Sequential Decision Model for Optimum Recognition," in *Biological Prototypes and Synthetic Systems,* Vol. I, edited by E. E. Bernard and M. R. Kare, Plenum Press, New York, 1962, pp. 270–277.

[19] R. A. Wiesen and E. H. Shuford, "Bayes Strategies as Adaptive Behavior," in *Biological Prototypes and Synthetic Systems,* Vol. I, edited by E. E. Bernard and M. R. Kare, Plenum Press, New York, 1962, pp. 303–310.

the allowable range of each variable.[20] With regard to the statistical decision procedures, optimal rules for decision are derived for certain classes of environments and for specific criteria. The problem is "solved" only if the experienced environment falls into a suitable class and the goal can be expressed by one of the available criteria.[21, 22] Considerable interest has been shown in nonparametric statistical inference; however, here again stationarity of environment is assumed.[23] A wide literature exists concerning optimization techniques. Some of these techniques may be viewed as being largely heuristic in that the solution is constructed on the basis of the demonstrated worth of methods in solving simpler problems.[24, 25]

Although heuristic modeling has been applied to a relatively wide range of human decision-making activities, it would appear constrained to remain within the capability of the human operator. Thus far, it has been difficult to realize, through this procedure, any new insight into the fundamental nature of intellect. The performance of each specific program is limited in "skill" to that of the programmer and the subject being modeled (even though a program's performance may demonstrate far greater speed and accuracy). In situations where the machine is allowed to play against other similarly programmed machines (or against itself), its ability may be further improved; however, this results only from the increase of experience, and *not* from a gain of new insight into the essential organization of intellect. *Artificial intelligence is realized only if an inanimate machine can solve problems that have, thus far, resisted solution by man; not because of the machine's sheer speed and accuracy, but because it can discover for itself new techniques for solving the problem at hand.*

A third approach toward artificial intelligence is possible. From a less egocentric standpoint, the human animal may be viewed as but a single artifact of the natural experiment called evolution. Though, certainly, man is an intelligent creature, there is no reason to believe

[20] S. S. L. Chang, *Syntheses of Optimal Control Systems,* McGraw-Hill, New York, 1961.

[21] A. Wald, *Statistical Decision Functions,* John Wiley and Sons, New York, 1950.

[22] H. Raiffa and R. Schlaifer, "Applied Statistical Decision Theory," Division of Research, Graduate School of Business Administration, Harvard University, Boston, 1961.

[23] S. Siegel, *Nonparametric Statistics for the Behavioral Sciences,* McGraw-Hill, New York, 1956.

[24] R. Bellman, *Dynamic Programming,* Princeton University Press, Princeton, New Jersey, 1957.

[25] G. Leitmann, *Optimization Techniques with Application to Aerospace Systems,* Academic Press, New York, 1962.

that he is the most intelligent creature that could possibly exist. Nature's experiment is continuing, and, in view of its past success in generating creatures of successively greater intellect, it would appear quite reasonable to anticipate that some future creatures will possess far greater intelligence than contemporary man. In this light, it would appear worthwhile to replace the process of modeling man as he now exists with the process of modeling evolution. This process would be carried out in the hope of reaching a greater understanding of the essential properties of intellect and, through this insight, gaining means for the synthesis of machines that display greater intelligence than has thus far been found in nature.[26]

In a sense, the evolutionary approach asks, "What might intellect be like some time in the distant future?" Although this question implies an outgrowth from man, the inquiry is intended to be far more general. For instance, there is no reason to believe that logical programs which result from modeling the process of evolution will bear any resemblance to the intellect of the actual progeny of contemporary man. (And yet, if the modeling of evolution can open the door to the design of greater intellect, it might also offer a significant step toward the solution of some of the important problems which face mankind today and, thereby, provide an indirect influence on the outcome of the continuing process of natural evolution.)

Modeling the natural process of evolution must be done with caution. As far as is known, this "experiment" of nature has occurred only once. The observed variability makes it difficult to believe that if the experiment were rerun, the outcome would be the same. The domain of possibilities is so great that it is highly unlikely that the human animal would be recreated even in a large number of similar evolutionary experiments. Thus, there appears to be an essential difficulty in replicating the evolutionary process. The investigator must identify those aspects of nature he considers to be fundamentally invariant properties of the environment; that is, properties that remain unaffected by the evolving creatures. In making this distinction, the investigator exercises his "freedom of design" through the selection of a set of underlying assumptions. On this basis, he may proceed to inquire into what evolution could be like: What kind of process will maximize the probability of success of those creatures that remain with respect to the challenge of their environment?

[26] L. J. Fogel, A. J. Owens, and M. J. Walsh, "Artificial Intelligence Through a Simulation of Evolution," in *Biophysics and Cybernetic Systems,* edited by M. Maxfield, A. Callahan, and L. J. Fogel, Spartan Books, Washington, D. C., 1965.

Both the bionic and heuristic programming approaches to artificial intelligence are primarily descriptive; that is, there is an attempt to model nature as it is observed to exist. In contradistinction, the evolutionary approach is primarily normative, in that it is an attempt to model evolutionary processes as they *might* occur in nature; to describe what *ought to be* rather than what *is*. The scientist does to know, while the engineer knows to do. In the evolutionary approach, the investigator takes an engineering attitude in his conceptual synthesis of the model.

It is recognized that there is no sharp distinction between descriptive and normative science. Descriptive science always includes an effect resulting from the orientation of the investigator in his search for what appears worthwhile to him. This is an ever-present normative influence. In a similar manner, normative science is carried out by an investigator who can understand nature only in terms that relate to his previous knowledge. Thus, his attempts to understand the limitations of what might exist in nature are always flavored by his descriptive outlook.

Various "organisms" might be used in the modeling of evolutionary processes; an increase in their intrinsic complexity offers, in turn, the benefit of an increase in the probability of the modeling being a success. (See Chapters 2 and 6.) But even after considering various combinations of mutation, selection, and recombination of these organisms, additional advantage may still be gained by injecting logical concepts derived from higher levels of abstraction. By analogy, model organisms may "move" on wheels, even though nature has not yet "invented" the wheel as a means for transport of living species.

Normative science specifically encourages the introduction of, and benefit from, optimal procedures whenever these are available and prove worthy of incorporation into the model. Similarly, models of decision-making processes may include optimal automated subroutines so as to benefit from the joint efforts of man and machine. Hopefully, the required decisions may be structured in terms of a hierarchy of subdecisions so that the mechanization of lower-level subdecisions allows the human operator to devote his full attention to higher-level tasks. It is expected that, as such man-machine-systems evolve, automatic subroutines will take their place at increasing levels within the hierarchy. Inquiry into methods for the worthwhile organization of decision-making automata will gain in recognition as fruits of the search for artificial intelligence materialize.

CHAPTER TWO

Evolutionary Programming

THE PREREQUISITE PREDICTION

Intelligent behavior may be viewed as a composite of ability to predict one's environment coupled with a translation of each prediction into a suitable response in the light of the given goal.[1] The ability to predict the environment with a reasonable degree of success is a prerequisite for intelligent behavior. For clarity of presentation, the following discussion will be concerned primarily with the problem of prediction. The term "environment" may include any source of information. For the sake of simplicity, and without sacrificing generality, the environment will be viewed as a sequential source of symbols taken from a finite alphabet. More specifically, the problem, at each point in time, is to devise an algorithm that will operate on the sequence of symbols thus far observed in such a manner as to produce an output symbol that is likely to agree with the next symbol to emerge from the environment. Evolutionary programming offers a means toward this end.[2]

In essence, a program is written to carry out a fast-time simulation of some of the fundamental aspects of natural evolution, at each point retaining that organism which can best cope with the given task. The organism is expressed in the form of a mathematical entity that describes a particular logic for transforming a sequence of input symbols into a sequence of output symbols. A "parent" organism is scored in terms of its ability to accomplish the desired decision-making on the

[1] At this point, the goal will be presumed to be well defined and given as dictum. A more detailed consideration of the nature and generation of goals is deferred to a more appropriate point in the discussion.

[2] See Appendix *A* for a procedure of particular interest as an explication of introspection.

basis of evidence at hand. This organism is mutated to yield an "offspring" which is then given the same task and scored in a similar manner. That organism which demonstrated the greatest ability to perform the required function is retained to serve as parent for a new offspring. Thus, nonregressive evolution proceeds to find better and better programs for attacking the problem at hand. At some point in real-time or whenever a sufficiently good "program-machine" has been found, an actual decision is made. Each time new information is received, an evolutionary search for a logic appropriate to the required decision-making is conducted. A more detailed account of the principles of evolutionary programming will now be offered.

To simulate evolution, it is necessary to choose a mathematical representation for the organism. Let the organism take the form of a finite-state machine, a machine defined in terms of a finite alphabet of possible input symbols, a finite alphabet of possible output symbols, and some finite number of possible different internal states. To specify such a machine, each of these states must be described in terms of those symbols that would emerge from the machine when the machine is in that state and receiving each of the possible input symbols. For example, a three-state machine is shown in Figure 2.1. The alphabet of input symbols is comprised of 0 and 1, whereas the alphabet of output symbols is comprised of α, β, and γ (input symbols are shown to the left of the virgule, output symbols are shown to the right). As shown in the "state diagram," an input of 1 to the machine while it is in state B will cause the machine to output an α and change

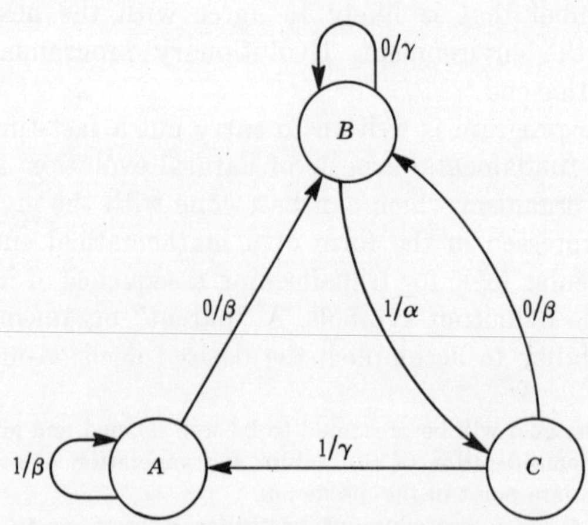

Fig. 2.1. A finite-state machine.

its internal state from B to C. On the other hand, an input of the symbol 0 when the machine is in state B causes an output symbol γ and the machine remains in state B. Thus, a sequence of input symbols is transformed into a sequence of output symbols as shown in Table 2.1, assuming that the machine is in state C when the first input is received. In general, the output sequence would have been different had the machine been in another "initial state." A finite-state machine, therefore, is completely specified by the state diagram and an identification of its initial state.[3] It is assumed that the machine acts only when an input symbol is received and that the operation is always completed before the next input symbol is received.

Table 2.1

Present State	C	B	C	A	A	B
Input Symbol	0	1	1	1	0	1
Next State	B	C	A	A	B	C
Output Symbol	β	α	γ	β	β	α

Suppose that it is desired to predict each next symbol in the input sequence. More specifically stated, at each point in real-time, the object is to devise an algorithm which will operate on the sequence of symbols thus far observed, in order to produce an output symbol that is likely to agree with the next symbol to emerge from the sensed environment. To accomplish this end, an arbitrary finite-state machine, M_0, is exposed to the sequence of symbols which have thus far emerged from the environment. (See Figure 2.2.) As this occurs, each output symbol from the machine is compared with the next input symbol. The percent correct score is a measure of the ability of this machine to predict the already experienced environment on the basis of the preceding symbols. (See Table 2.2.) The first row of this table indicates the sequence of state as the machine responds to the sequence of input symbols shown on the second row. State B was arbitrarily taken to be the initial state. The most recently experienced symbol is shown to the immediate left of the vertical line while the symbol to the immediate right of this line is the symbol to be predicted. The third row shows the sequence of symbols resulting from the transduction. Note that these symbols are shown one position to the right in

[3] That state the machine is in when it receives the first symbol of the input sequence under consideration.

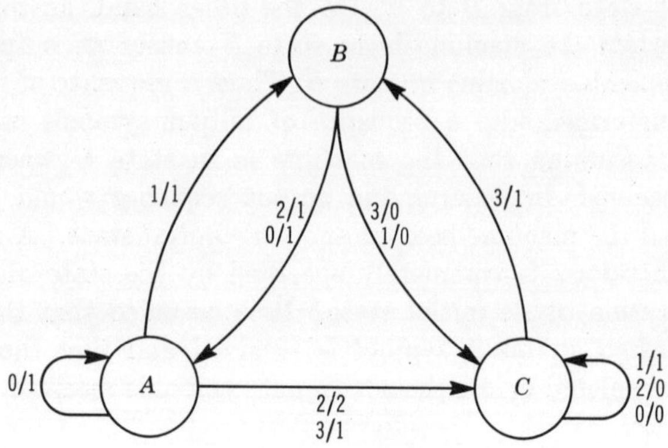

Fig. 2.2. Machine M_0.

order to facilitate comparison. The output symbol shown to the right of the vertical line is this machine's prediction of the unknown symbol.

The last row is generated by referencing the goal which is expressed in terms of an error cost matrix, $A = [a_{ij}]$. For this example, $a_{ij} = 0$ if $i = j$, and $a_{ij} = 1$ if $i \neq j$. The cost of each transduction is found from the error-cost matrix; i referencing the machine output, and j referencing the next symbol of the input sequence. A

Table 2.2

Present State	*B*	*A*	*C*	*C*	*C*	*C*	*B*	*C*	*C*	*B*	*A*	*B*
Input Symbol	2	2	1	0	1	3	3	0	3	0	1	2
Output Symbol		1	2	1	0	1	1	0	0	1	1	1
Error Cost		1	1	1	1	1	1	0	1	1	0	

measure, P_1, of the ability of the machine to predict each next symbol is calculated by summing the cost of the error and dividing this sum by the number of terms. Thus, the average cost of the errors of machine M_0 is found to be

$$P_1 = \frac{\text{Sum of the Error Costs}}{\text{Number of Terms}} = \frac{8}{10} = 0.8$$

An offspring of this machine is then produced through mutation; that is, through a single modification of the parent machine in accordance with some mutation noise distribution. The mode of mutation is determined by the interval within which a number selected from a random number table lies. The intervals are chosen in accordance with a probability distribution over the permitted modes of mutation. Additional numbers are then selected in order to determine the specific details of the mutation. Thus, the offspring is made to differ from its parent either by an output symbol,[4] a state transition, the number of states, or the initial state. For example, the offspring of M_0, M_1, might appear as in Figure 2.3, where the two differing output symbols were set deterministically.

Table 2.3 indicates the evaluation of the prediction capability of this machine over the same sequence of experience (this being termed

Table 2.3

Present State	*B*	*A*	*C*	*C*	*C*	*C*	*B*	*C*	*C*	*B*	*A*	*B*
Input Symbol	2	2	1	0	1	3	3	0	3	0	1	2
Output Symbol		1	2	1	0	1	3	0	0	3	1	2
Error Cost		1	1	1	1	1	0	0	1	1	0	

the "recall"). The resulting average cost of the errors, P_1, is found to be 0.7. This offspring demonstrates a superiority over its parent with respect to the given task; hence the original parent, M_0, is dis-

[4] In the case of a 2-symbol environment, a deterministic procedure can be used to replace this type of mutation. As each symbol from the environment is predicted on the basis of the preceding symbols, score is maintained of the relative frequency of success of each state transition. A predictive-fit score of greater than 0.5 can then be ensured by the reversal of output symbols on those state transitions which were "more often wrong than right."

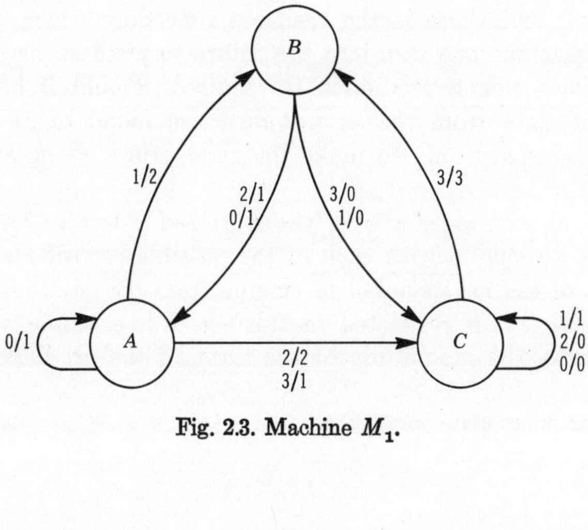

Fig. 2.3. Machine M_1.

carded and the offspring, M_1, is retained to serve as a parent. If machine M_1 had a predictive error in excess of 0.8, it would have been discarded, and the next evaluated offspring would be generated by mutating machine M_0.

At any point in time, the remaining machine can be used for actual prediction. As seen in Table 2.2, the machine M_0 had predicted that the next symbol to emerge from the environment would be a 1. This was incorrect. Table 2.3 indicates that the machine M_1 correctly predicted the next symbol to emerge from the environment. The risk associated with each prediction is roughly related to the average cost of the errors in "predicting" the past.

Machine M_1 is now used to parent succeeding offspring, these being evaluated over the same recall. Thus, the evolution continues in fast-time in preparation for an actual prediction. Such prediction may take place periodically, aperiodically, or upon request. It may be made whenever the average error in predicting the past has fallen below a predetermined value, when some prechosen number of offspring have been evaluated, or when an appropriate number of generations[5] have occurred. Of course, in general, the longer the time interval before each prediction, the greater the expectation of success. Similarly, the greater the speed of the computer facility (increase in the number of evaluated offspring), or the larger the available memory (increase of their permissible size), the greater the evolutionary prediction capability.

The process continues in this same manner. After each actual prediction, the remaining machine serves to generate successive offspring. In essence, the evolutionary program provides two processes: There is the iterative mutation and selection of machines (a continual search for that decision logic which would have been best had it been used in the past), and there is the read-out procedure which carries the remaining machine one step into the future to yield an actual prediction whenever such a prediction is required. Should it happen that some of the data from the environment are found to have been in error it is necessary only to make the appropriate changes to correct the recall.

The goal, as mentioned above, was expressed in terms of a particular set of costs associated with each of the possible correct and incorrect predictions of the next symbol to emerge from the environment. Note that the goal is not restricted to this set. For example, if the goal is to minimize the magnitude of the error of the prediction of each

[5] Defined at an appropriate point below.

next symbol, the elements of the error cost matrix would take the values $a_{ij} = |i - j|$. If it is desired to minimize the mean squared error of such prediction, the elements would become $a_{ij} = (i - j)^2$. Or, if "a miss is as good as a mile," the error matrix should have equal nonzero off-diagonal terms and zero on the diagonal. In fact, if there is some greater worth associated with correctly predicting a certain symbol, this can be reflected in the cost matrix expression of the goal. For example, the goal expressed in Table 2.4 indicates that it is most important to predict properly the symbol 0, less important to predict correctly the symbol 1, and least important to predict correctly the symbol 2. At the same time, this goal expresses the fact that it is more desirable to err by predicting a 1 when the actual symbol is a 0 as opposed to predicting a 0 when the actual symbol is a 1. The relative worth of each possible outcome is expressed in such a statement of goal. And there is no additional cost in using such a complex goal. As each output symbol of the machine under evaluation is compared to the next symbol which actually emerged from the environment, reference is made to the given goal matrix. The sum of the associated costs provides the desired measure of the worth of the particular machine in terms of its ability to predict the sequence of symbols already experienced. In general, then, the sequence of machines that will evolve will be a function of the goal. Further, there is no need to restrict the evolution to a single invariant goal. If the goal is changed as real-time experience proceeds, the sequence of retained machines will gradually reflect the change of goal. Of course, all that has been said concerning goals expressed in the form of penalty matrices applies equally well if the goal is expressed in the form of payoff matrices.

Table 2.4

		Predicted		
		0	1	2
Actual	0	0	3	4
	1	4	1	4
	2	4	4	2

The evolutionary technique offers even greater versatility. For example, the desire to predict each *second* symbol in the future can be satisfied simply by scoring each offspring in terms of the correspondence between its output symbols and those symbols which emerge from the environment two symbols later. By the same token, appropriate scoring of the offspring permits the prediction of any particular future symbol, the average of some set of future symbols, or, indeed, any well-defined function over the future.

Note that the evolutionary program, as just described, is suitable for the prediction of multivariate environments. The symbols of the sensed parameters of the environment may be combined into a new alphabet of symbols that describes the environment in an unambiguous manner. The resulting predictions are made in the same alphabet so that they may be interpreted in terms of separate predictions for each of the parameters. Of course, the number of symbols in such a language rises rapidly as a function of the number of variables. However, in principle, there is no restriction.[6]

MORE GENERAL GOAL-SEEKING

But the purpose of evolutionary programming need not be restricted to prediction in any sense. For example, problems of interdiction may be addressed by comparing the input variable derived from the environment with another parameter of the environment that exists concurrently but can only be measured in retrospect. Here is the case of *diagnosis*. A number of parameters of the system under consideration are measured. The intent is to determine some other parameter which is not open to immediate measurement.

Note that in the case of interdiction the input and output languages of the evolving machines might be different. The input symbols may be individually associated with a set of possible stimuli and output symbols with a set of alternative responses, the goal being to optimize the performance with respect to some well-defined evaluation of the response. A distinction need no longer be made between prediction and the translation of prediction into response. The evolutionary program recommends each action in the light of its expected overall worth with respect to the given goal.

In the broadest generality, a goal is completely specified only if it includes a statement of the relative importance to the decision-maker of each point in future time and there is a complete penalty or payoff matrix expressed for each point in the future. To this point, it has been tacitly assumed that the importance function of the goal-seeking entity has been restricted to a single point in its future time. That is to say, if the concern is to predict the next symbol no importance is associated with the second symbol into the future, or, for

[6] Reference is made to Chapter 4 which contains examples of this technique.

that matter, with any other symbol farther into the future. In principle, it is possible to accept a more complex importance-weighted expression of goal, in that each offspring can be made to continue yielding its prediction over the indefinite future. Each of these can be translated into their relative worth. Once the actual environment becomes known, the evaluation of the logic is then found by combining these metrics by means of relative importance weighting. Note the importance of having available actual information concerning the environment. The prediction of sequences of symbols assumes interest in situations wherein measurements of the environment can be made only at certain times. In its most general form, evolutionary programming provides a means for seeking any well-defined goal in a manner consistent with the allowable costs. Each offspring is evaluated in terms of its own total worth as an organism . . . the value of its sequence of actions in the light of previous circumstance is degraded by the cost of its very existence.

Efficiency of the simulated evolution can be improved in a number of ways. For example, any available information concerning the underlying logic of the environment can be used effectively by translating this logic into the form of a finite-state machine that would express a proper logic for the desired decision-making under such environment. This machine may then serve as the initial machine. If this hint is reasonably correct, the evolution may require fewer generations to attain the same score; if it is incorrect, this introduction of false information in no way precludes solution of the problem, although it may be expected to reduce the efficiency of the procedure.

By the same token, the evolutionary technique permits the examination of tentative results at any point during the process. If a sufficiently worthwhile design has not yet evolved, the evolution may be reinstituted, using as the initial machine the previously resulting machine. In this manner, the procedure may be extended until some suitable goal-directed complexity-costed organism has evolved or the available expenditure has been exhausted.

The maxim of parsimony may be directly incorporated into the evolutionary procedure by increasing the penalty of each machine by a measure of its complexity. The amount of this penalty may be influenced by the particular characteristics of the computer facility upon which the simulation is to be carried out. For example, the penalty for complexity might take the form of a constant times the number of states in the machine under evaluation. If the available memory is severely restricted, it might be desirable to use a constant multiplied by the square of the number of states, or some other suitable expression. In any case, the procedure should restrict the offspring from growing to size beyond the available memory of a computer. In general, the cost for complexity may take any form as long as its value may be directly computed from the specification of the machine. (See Appendix B.)

If the environment is primarily statistical in nature, it is reasonable to suspect that a smaller machine will provide a suitable representation for the decision logic. Such a machine may characterize conditional dependencies and because of its small size have considerable statistical validity associated with each transition. On the other hand, such a small machine might present an oversimplified "view" of a complex deterministic environment. It becomes evident that the penalty for complexity provides a parameter which can be adjusted over a scale ranging from a purely statistical view to a purely deterministic view of the environment. The investigator can have the selection of this parameter at his disposal. However, if his decision is made on insufficient grounds, he may have imposed an unnecessary restriction on the evolutionary capability. It is, of course, possible to provide a procedure that will continually optimize the cost for complexity as the program is exercised on the environment. For example, this procedure might include the comparative evaluation of a pair of machines prior to the generation of each offspring. The "functional-fit" average cost of the error indicates which of these machines is most suitable; and reference to its penalty for complexity, as compared with that for the other machine, furnishes a basis for its suitable modification. Thus, as the evolutionary program proceeds, it should continually revise its opinion as to the nature of the environment, this ranging over a scale from purely statistical to purely deterministic.

At each point in time, therefore, the evolutionary technique provides a nonregressive search through a domain of finite-state machines under the constraints imposed by the available computation capability for that machine which is most likely to achieve the given goal. This search may be viewed as a selective random walk, a hill-climbing procedure in a hyperspace defined to include the finite-state machines, and an additional coordinate on which is measured the average cost of the error score. The danger of becoming trapped on a secondary peak can be overcome by permitting multiple mutation, with the multiplicity being a function of the difference in the average error score of successive generations. Thus, as the search nears a peak, greater and greater attention may be devoted to generating more radical offspring in the hope of striking a point which may lie higher on the slope of another peak.

In general, a suitable choice of the mutation noise can increase the efficiency of evolution. For example, an increase in the probability of adding a state generates a wider selection of larger machines, and this should benefit evolution against a complex environment. In fact, the probability distribution over the modes of mutation can be made to depend upon the evidence acquired within the evolutionary process itself. Thus, an experienced greater relative frequency of success for, say, changing the initial state might be made to increase the probability of this mode of mutation.

Evolutionary programming can be used even in the face of nonstationary environments because it provides an iterative search for a "best" logic. But selection of only the best logic may be an overly severe policy. Certainly, all those offspring that, in their evaluation over the recall, demonstrate a significant decision capability characterize the environment in some meaningful manner. Why not mimic natural evolution and save the best few machines at each point in time? In general, the best offspring is most likely to produce a superior offspring, and thus it should receive most attention in terms of mutative reproduction. Lower-ranked offspring may be regarded as insurance against gross nonstationarity of the environment. The distribution of mutative effort may well be in proportion to the normalized evaluation scores. Evaluated offspring are inserted into the rank-order table of retained offspring, and a *generation* is said to occur whenever an offspring is found which has a score equal to, or greater than, the score of the best machine. All of the retained machines need not lie on the slopes of the peak identified by the best machine. Thus, the saving of the best offspring may maintain a "cognizance" over several peaks, with the relative search effort being distributed in proportion to the expectation of significant new discoveries.

The recombination of individuals of opposite sex appears to benefit natural evolution. By analogy, why not retain worthwhile traits that have survived separate evaluation by combining the best surviving machines through some genetic rule; mutating the product to yield offspring? Note that there is no need to restrict this mating to the best two surviving "individuals." In fact, the most obvious genetic rule, majority logic, only becomes meaningful with the combination of more than two machines.

It is always possible to draw a single state-diagram which expresses the majority logic of an array of finite-state machines. Each state of the majority logic machine is the composite of a state from each of the original machines. Thus, the majority machine may have a number of states as great as the product of the number of states in the original machines. Each transition of the majority machine is described by that input symbol which caused the respective transition in the original machines, and by that output symbol which results from the majority element logic being applied to the output symbols from each of the original machines. To illustrate, Figure 2.4 indicates three original machines and their individual probabilities of success, p_i. The initial state of each machine is indicated by a short arrow pointing to that state. The output of these machines is to be combined through a majority logic element that weights the importance of each output by the demonstrated probability of success of that machine. The resulting majority machine is shown in Figure 2.5.

As long as there are only two original machines, the weighted majority logic machine reduces to the better of these two machines. A more interesting situation occurs when there are three original machines, these being rank ordered so that $p_1 \geq p_2 \geq p_3$. If the environment is

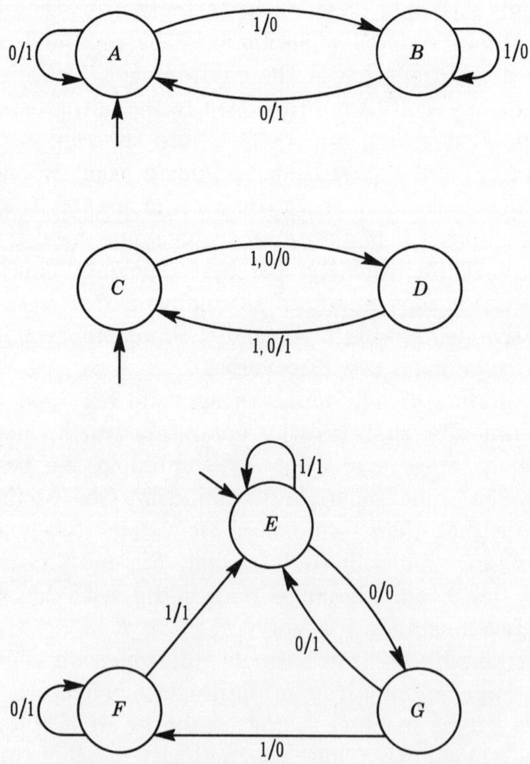

Fig. 2.4. Individual machines to be combined.

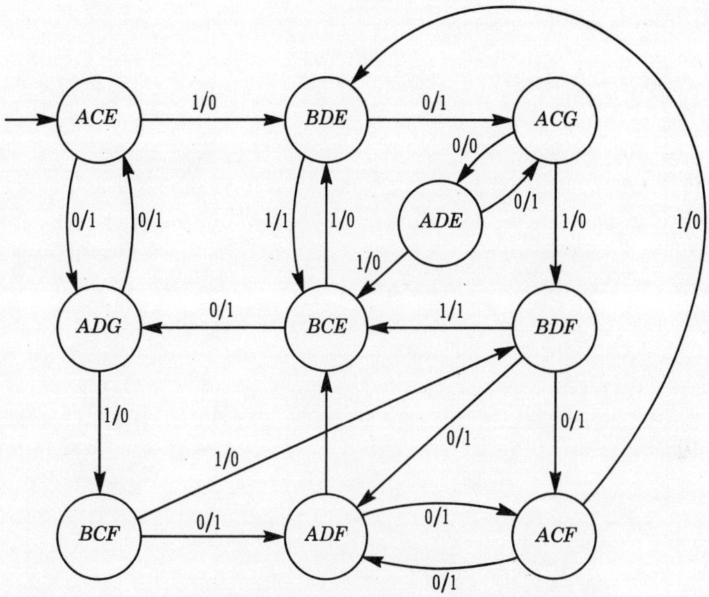

Fig. 2.5. The resulting majority logic machine.

described in terms of only two symbols these scores cannot be less than 0.5 in view of the already described deterministic output symbol reversing technique; therefore, it will always be true that $p_1 \leq p_2 + p_3$. Because of this inequality, the weighted majority machine reduces to the simple unweighted majority machine.

The score for the majority machine, p_M, can be expressed as the sum of the ways in which success can be attained, that is,

$$p_M = p_1p_2p_3 + p_1p_2q_3 + p_1q_2p_3 + q_1p_2p_3$$

where q_i is the probability of failure, so that $p_i + q_i = 1$. This expression can be simplified to

$$p_M = p_2p_3(1 - 2p_1) + p_1(p_2 + p_3)$$

Now the majority machine is only of value if $p_M \geq p_1$; that is, if $p_M - p_1 \geq 0$. Substituting yields the inequality

$$p_2p_3(1 - 2p_1) + p_1(p_2 + p_3) - p_1 \geq 0$$

This relationship can be expressed in graphical form for specific values of p_1 and solved for maximum value of p_M. Figure 2.6 shows

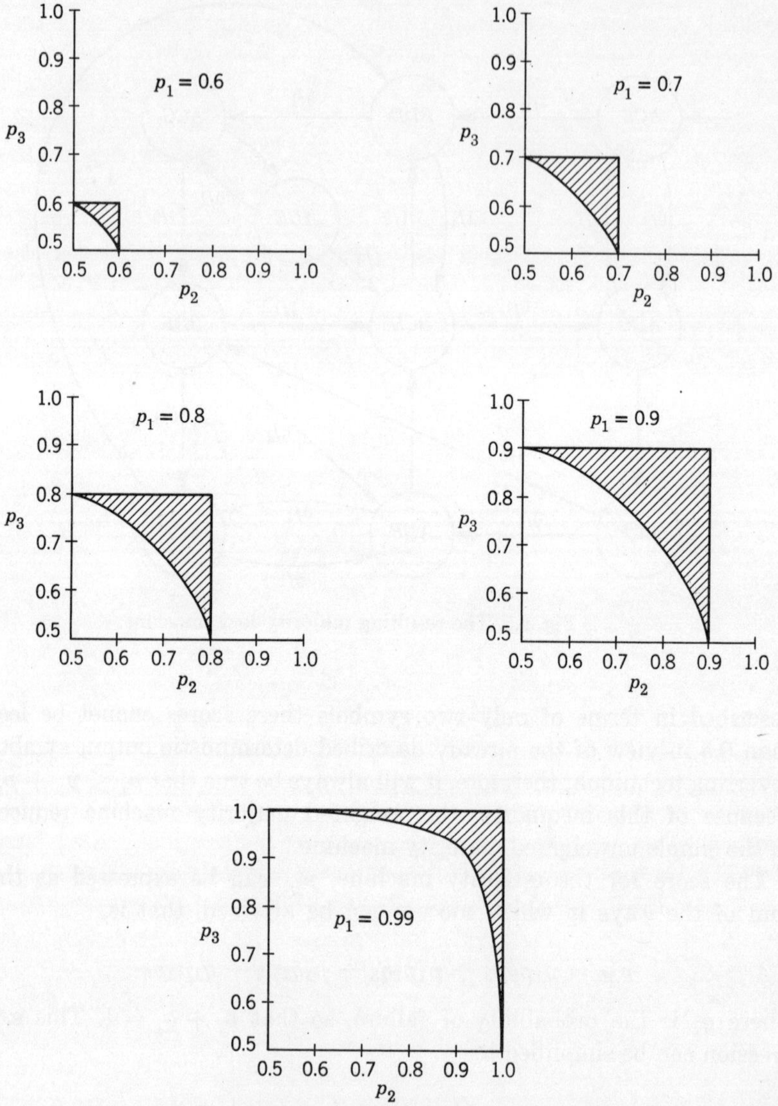

Fig. 2.6. Majority logic improvement domains.

the regions in which pairs of values of p_2 and p_3 yield a $p_M \geq p_1$ for these values of p_1. Table 2.5 indicates the corresponding values for the maximum p_M.

This same procedure can be used for the evaluation of the benefit to be achieved through a majority logic element over any number of original machines; however, the required calculations rapidly becomes cumbersome. Further, for large numbers of original machines, it becomes worthwhile to examine the behavior of the weighted majority logic machine as opposed to the simple majority logic machine. Clearly, this opens the door to many new possibilities. For example, it might prove fruitful to explore the combining of the best machines of several different generations in the hope of finding a model of the models which have thus far proven to be most successful.

Table 2.5

The Maximum p_M Related to Specific p_1

p_1	Maximum p_M
0.6	0.648
0.7	0.784
0.9	0.972
0.99	0.9997

Up to this point, it has been tacitly assumed that each offspring is to be evaluated over the entire available experience. Obviously, this is not always the best policy. For example, if the environment changes its statistical properties it may be better to restrict the recall to some past sequence of symbols in order to enhance the probability of finding a most suitable decision logic for the present. In the face of complete ignorance concerning the logical construction of the environment, the choice of a particular length of recall might prove disastrous. However, it is possible to use a procedure similar to that suggested earlier for determining successively more appropriate values for penalities for complexity. At each point in time, two or more machines are evolved over different recall lengths and their individual worth compared, resulting in a means for determining the next lengths of recall to be considered. The complete experience should be retained for as long as possible so that the recall length may, if desired, be allowed to grow. The problem becomes increasingly difficult with the restriction of the available memory to the point where the only alternative is "slide"; that is, keeping the recall length fixed.

All these procedures, intended to increase the efficiency of evolution, pose a danger if the environment is interactive. An intelligent adversary might discover the specific procedure and use this knowledge to construct an obverse strategy.

Significant difficulties are encountered in any attempt to express the principles of evolutionary programming in terms of conventional mathematics. The succession of finite-state machines considered is dependent partly upon the statistical process of mutation and partly upon the environment. Thus, any formal representation of the process would have to be restricted to a special class of environments. But the very point of evolutionary programming is its versatility. Here is a procedure that includes a random process in order to generate a never-ending sequence of alternative hypotheses concerning suitable logic and decision-making. Information from this source is selected in the light of the given goal and the experienced environment.

Classical techniques for decision-making are based on the presumption that the environment is either stationary or can be viewed as stationary through a simple transformation. Further, it is conventional to presume that the environment is independent of the decision-maker. Neither of these classical assumptions is required for evolutionary programming. Its value is achieved through the inheritance of worthwhile properties through the extensive search made possible by the advent of high-speed digital computers.

CHAPTER THREE

Prediction Experiments

2-SYMBOL PREDICTION EXPERIMENTS

A series of increasingly difficult prediction experiments was conducted in order to demonstrate the feasibility of evolutionary programming. In the interest of brevity, only some of these experiments will be reported.

An evolutionary program designed to accept and predict 2-symbol environments was written in Fortran II language for use on the IBM 7094 computer. (See Appendix *C*.) Unless otherwise noted, all these experiments were initiated with the same arbitrary five-state machine, starting in state 1. (See Table 3.1.) At each point in time, only the three best evolved machines were retained to serve as a basis for generating offspring. In these experiments, a maximum of five generations was permitted to occur prior to making each actual prediction.

Table 3.1

State	Input Symbol	Next State	Output Symbol
1	1	2	1
1	0	1	1
2	1	3	1
2	0	2	0
3	1	4	1
3	0	5	0
4	1	5	1
4	0	3	0
5	1	1	1
5	0	2	0

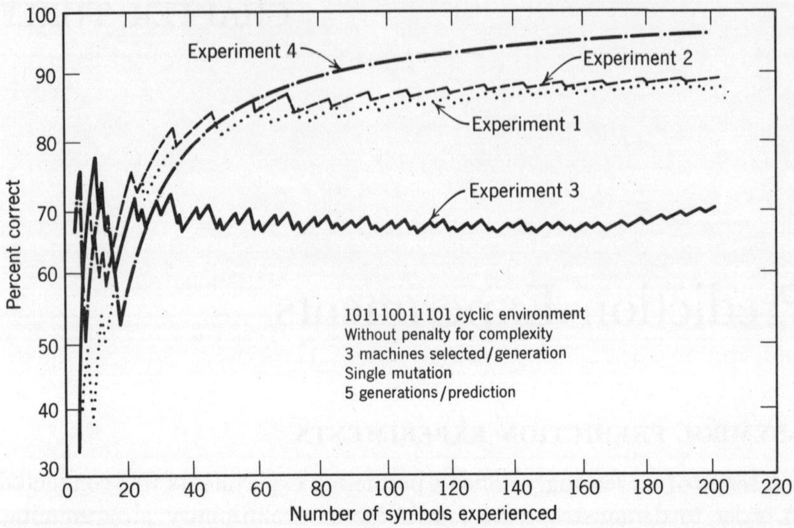

Fig. 3.1. A comparison of the prediction score for four different sequences of random numbers.

The first experiments were attempts to predict an environment generated by an unending repetition of the arbitrary sequence 101110011101. The first four experiments were conducted to examine the sensitivity of the prediction capability to the nature of the mutation noise. In these experiments, no penalty for complexity was imposed, and only single mutation was applied in order to generate each offspring. The mutation noise was uniformly distributed over five modes of mutation: add a state, delete a state, randomly change a next state, randomly change the initial state, and change the initial state to the second state the machine assumed under the available experience. The last of these was used to permit the possibility of "phase shifting." The sequence of random numbers used for each of these experiments was drawn from a generally accepted source.[1]

Figure 3.1 shows the results of these four experiments in terms of the percent correct as a function of the number of symbols experienced from the environment. In each of these experiments several thousand offspring were evaluated, and the predictor machines were generally of eight to ten states. In Experiment 4, a sequence of perfect predictor-machines was found after the nineteenth symbol of experience.

[1] *A Million Random Digits with 100,000 Nominal Deviates,* Rand Corporation, Free Press of Glencoe, New York, 1955.

Poorest prediction occurred in Experiment 3, but even this experiment demonstrates a distinct tendency for the prediction to improve after the first dozen repetitions of the basic pattern of the environment. The first experiment (resulting from random series 1) is considered typical. This experiment will be referenced as a basis for comparison.

The effect of imposing a penalty for complexity is shown in Figure 3.2. The solid curve is the result of Experiment 5 which was identical to Experiment 1 except that a penalty of 0.01 per state was imposed. Note that a significant difference of prediction capability is apparent only in the initial period. The benefit derived from use of such a penalty for complexity is evident from Figure 3.3 which indicates the significant reduction in the complexity of the evolved predictor-machines. It is interesting to note that perfect prediction of this repetitive environment can be achieved by various eight-state machines (three of which are shown in Figure 3.4). Restricting the number of states may preclude finding such machines.

With an evident need for larger predictor-machines it is reasonable to suspect that an increase in the probability of adding a state might improve the prediction capability. Experiment 6 was a repetition of Experiment 1 except for the mutation noise. In this experiment the probability of adding a state was increased from 0.2 to 0.3 (this in-

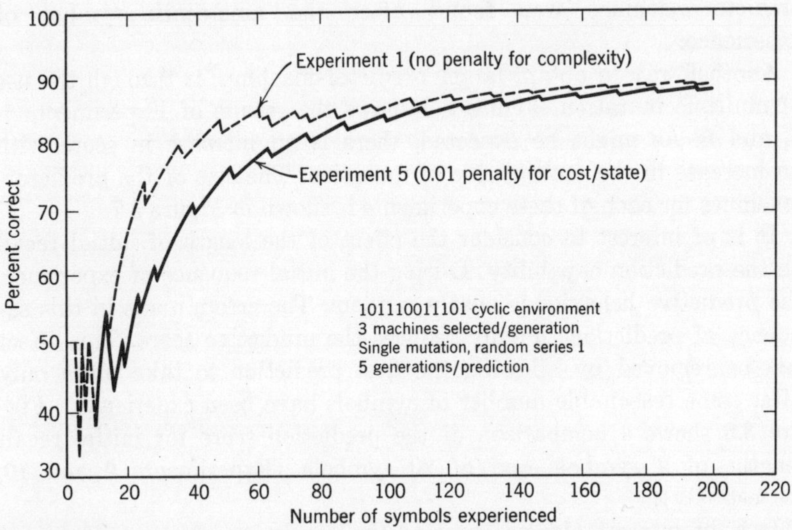

Fig. 3.2. A comparison of the prediction score with and without a penalty for complexity.

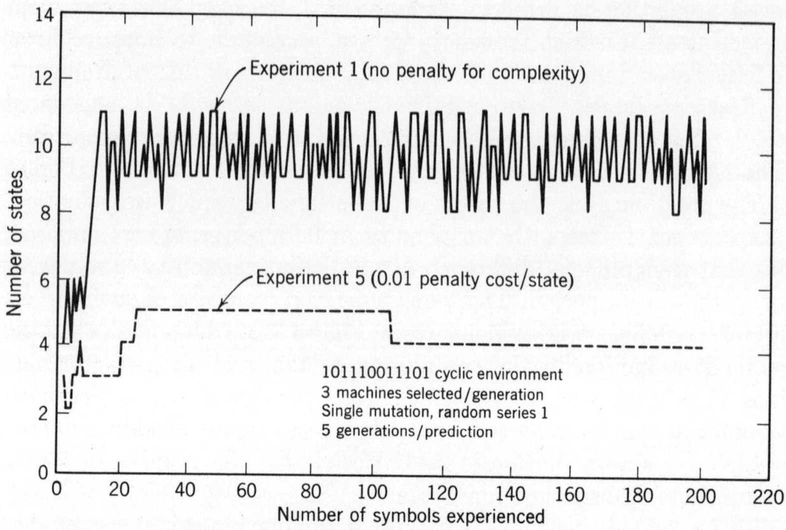

Fig. 3.3. A comparison of the number of states in the best predictor machines with and without penalty for complexity.

crease being balanced by a reduction in the probability of deleting a state). Figure 3.5 shows the benefit derived. A sequence of perfect predictor-machines was found after the nineteenth symbol of experience.

Another way to obtain larger predictor-machines is through the use of multiple mutation. Figure 3.6 shows the results of Experiments 1, 7, and 8. As might be expected, there is an increase in score with an increase in the multiplicity of mutation. The size of the predictor-machines for each of these experiments is shown in Figure 3.7.

It is of interest to consider the effect of the length of initial recall on the prediction capability. During the initial sequence of experience, the predictive behavior is nearly random. The errors made in this sequence of predictions unduly reduces the predictive score. This effect can be removed by allowing the first prediction to take place only after some reasonable number of symbols have been experienced. Figure 3.8 shows a comparison of the prediction score for initial recall lengths of 2 symbols and of 20 symbols (Experiments 9 and 10, respectively).

It is of interest also to examine the prediction of an environment which undergoes a sudden radical change in behavior. Figure 3.9 shows the result of Experiment 11 in which the environment consists of the

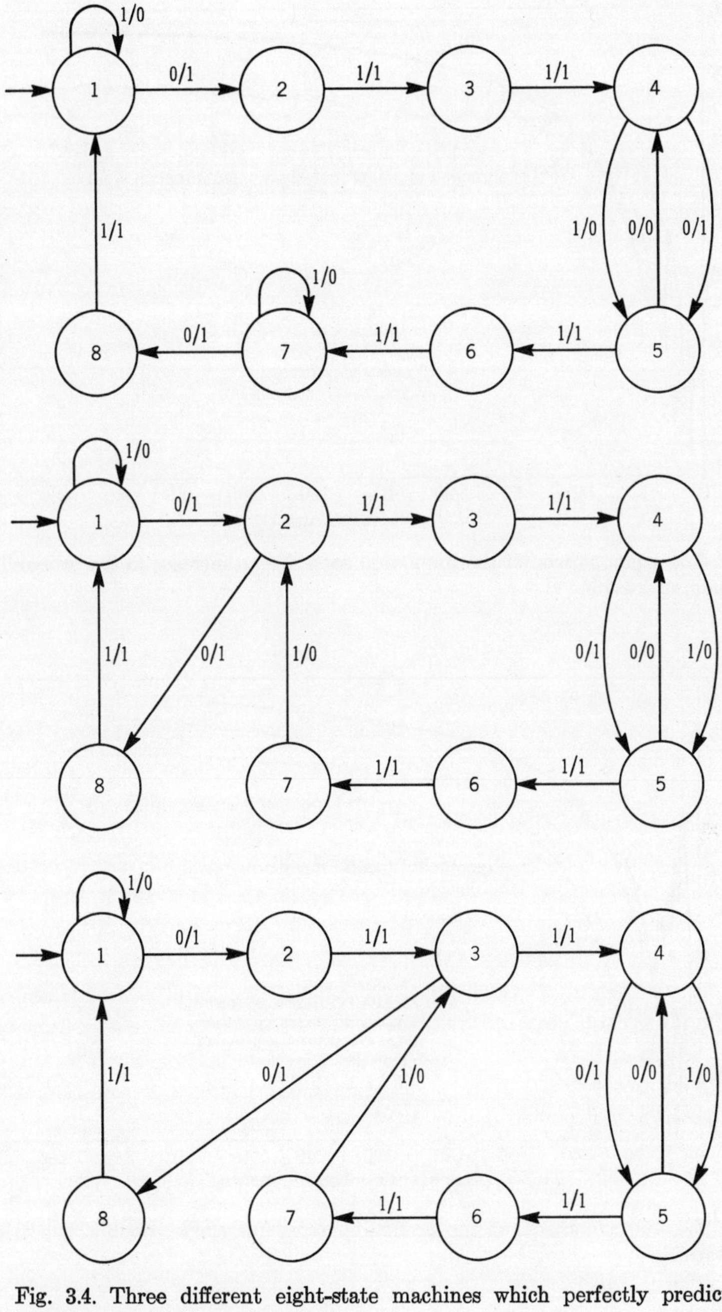

Fig. 3.4. Three different eight-state machines which perfectly predict 101110011101 cyclic environment.

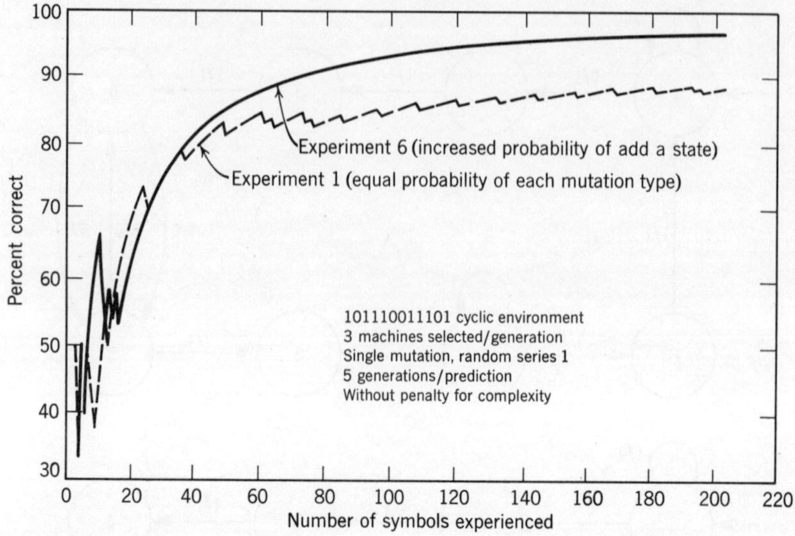

Fig. 3.5. A comparison of the prediction score for an increase in the probability of adding a state.

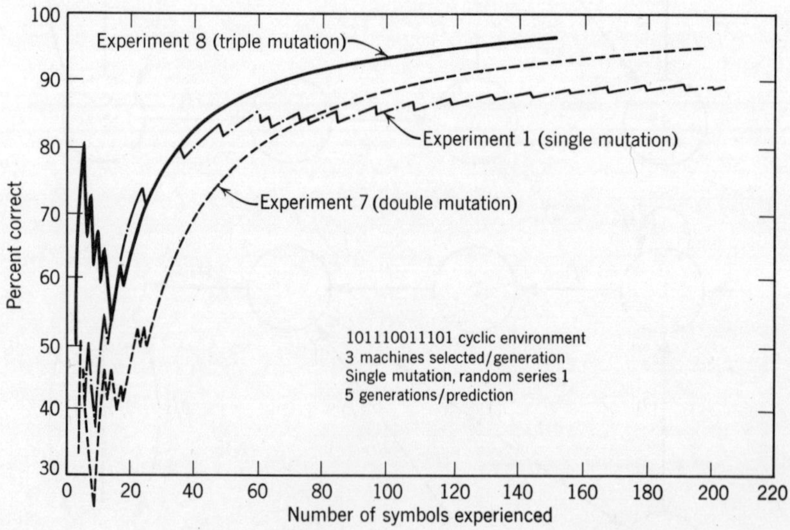

Fig. 3.6. A comparison of the prediction score for single, double, and triple mutation.

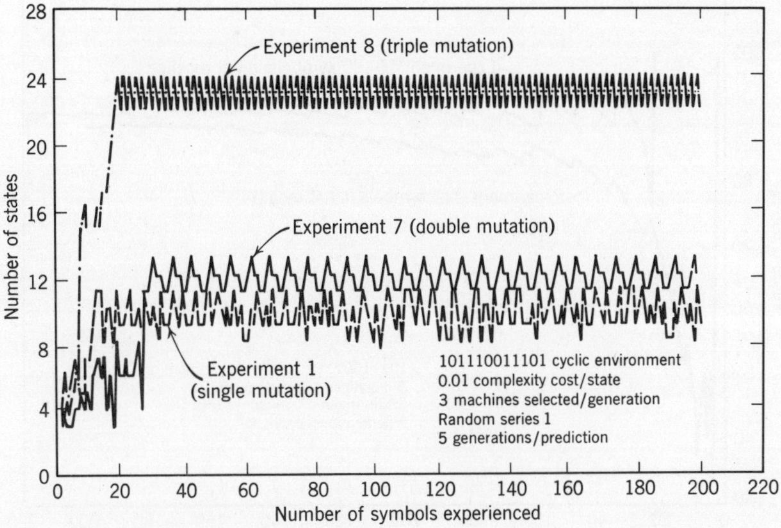

Fig. 3.7. A comparison of the number of states of the best predictor machines for single, double, and triple mutation.

sequence 101110011101 repeated ten times, this being followed by the inverse sequence 010001100010 repeated ten times. The initial recall is 20 symbols long, that is, the 21st symbol to emerge from the environment is the first symbol to be predicted. This initial recall permitted the selection of machines which correctly predicted the 21st, 22nd and 23rd symbols before an error was made. The prediction score then fluctuated around 0.9 until the first phase of the environment came to an end. At this point there occurred a series of errors, and the prediction score dropped to 0.814 by the 159th symbol. Thereafter, the selected machines exhibited "adaptation" and the prediction score again began to rise.

In order to properly portray this adaptation, it is fitting to replot the prediction score, beginning with the 141st symbol. The dashed curve on Figure 3.9 indicates the prediction score for the second phase of the environment but with the initial machines resulting from that phase of experience. It is evident that some "unlearning" had to take place. This second-phase prediction score compares favorably with that of the first phase when it is taken into account that the machines are measured in their predictive fit over the entire length of experience. This explains the rapid growth in the number of states of the machines used for prediction, as shown in Figure 3.10. During the first phase

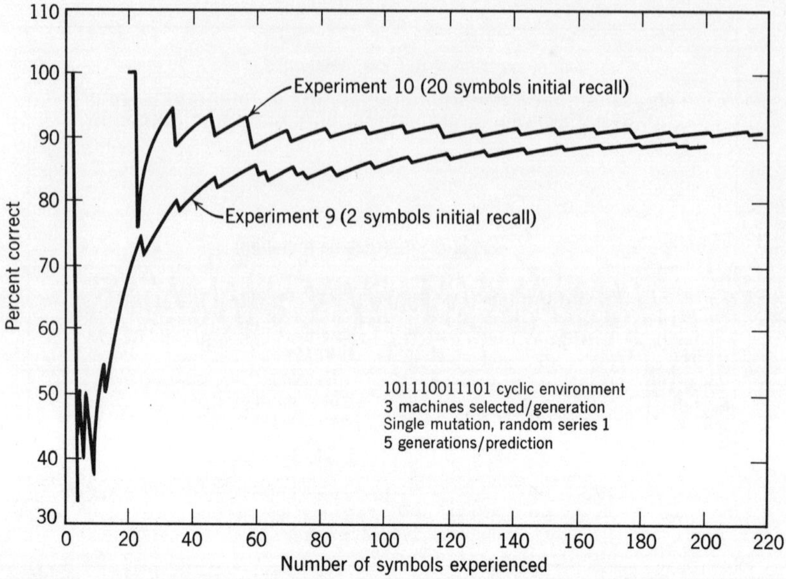

Fig. 3.8. A comparison of the prediction score for different lengths of initial recall.

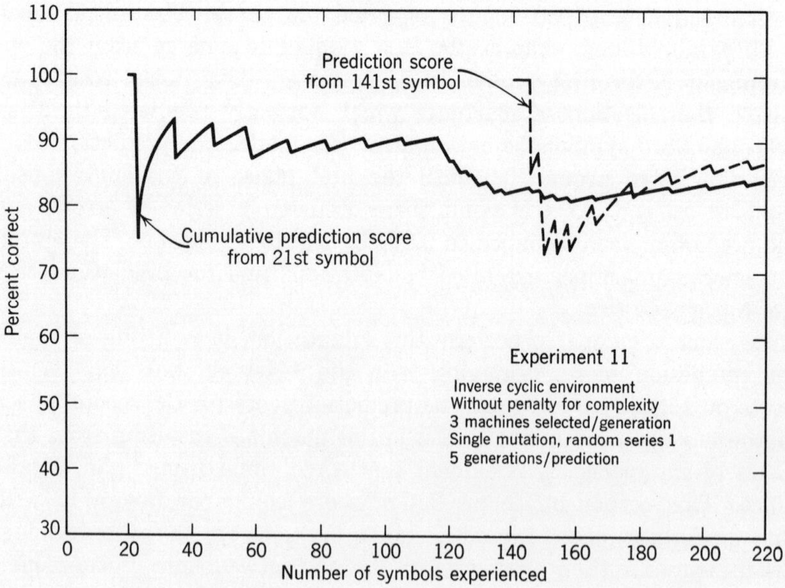

Fig. 3.9. A comparison of the prediction score for a reversing 12-symbol cyclic environment.

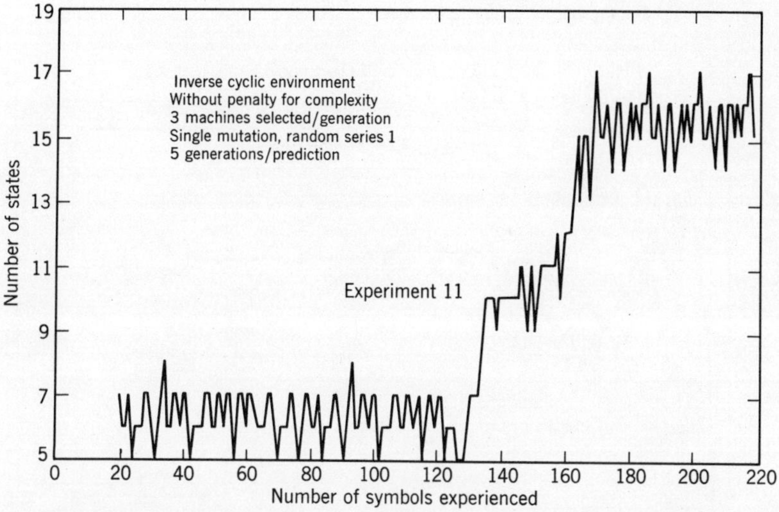

Fig. 3.10. The number of states in the best predictor machines for the reversing 12-symbol cyclic environment.

of the environment the number of states had remained near six, whereas in the second phase the number of states stabilized in the vicinity of sixteen.

A second level of difficulty is imposed by a statistically stationary environment that results from the repetitive flipping of a 0.7 weighted coin. With *a priori* knowledge of this fact, it is possible to predict this environment with an asymptotic score of 0.7. Figure 3.11 indicates the result of application of the evolutionary procedure with no such prior knowledge. The effect of three different lengths of recall are shown. It is apparent that too small a sample size degrades the prediction capability. Restriction of the recall to the last 13 symbols (Experiment 12) reduces the prediction capability (this length of recall being chosen to correspond with the length of experience upon reaching the first peak of average prediction score under continually growing recall). A recall of the last 50 symbols (Experiment 13) produces only slightly poorer results than no restriction in the length of recall, however, this constraint considerably improves the efficiency of the computation process by greatly reducing the computation time required for evaluation of each machine, especially in longer sequences of experience. Experiment 14 was performed with growth of recall. All three curves tend to approach an average score of 62 percent. Note that the best possible average prediction score of this environment would

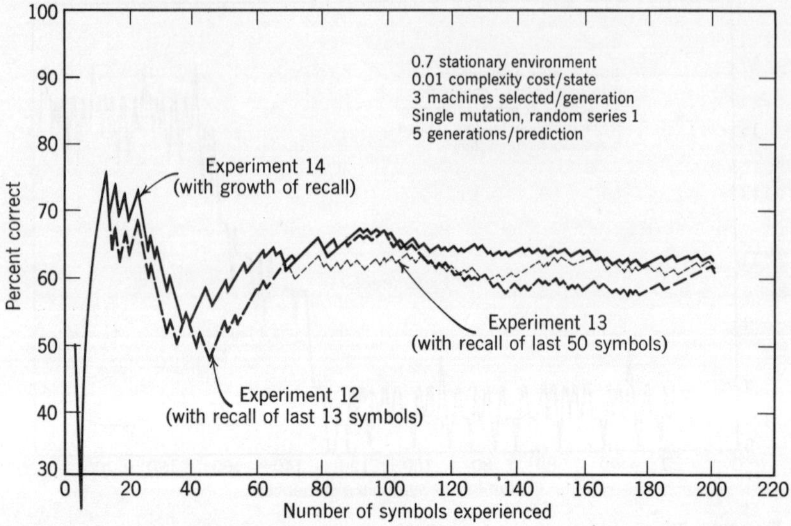

Fig. 3.11. A comparison of the prediction score for a 0.7 weighted stationary environment with various lengths of recall.

be 70 percent; however, to achieve this score, it would be necessary to use *a priori* information concerning the nature of the environment.

An even more difficult environment is the statistically nonstationary sequence of symbols generated by classifying each of the increasing possible integers as being prime (represented by the symbol 1) or nonprime (represented by the symbol 0). That is to say, the environment consists of the sequence 01101010001 . . . , wherein each symbol depicts the primeness of the numbers 1,2,3,4,5,6,7,8,9,10,11 . . . , respectively. Although in principle this deterministic environment should be perfectly predictable, no finite procedure is known for accomplishing such prediction. In the words of Hardy and Wright, "Indeed it is unlikely that such a formula is possible, for the distribution of the primes is quite unlike what we should expect on any such hypothesis."[2]

Figure 3.12 indicates the prediction score achieved against this nonstationary environment (Experiment 15). After the initial transient, the prediction score gradually increases to reach 78 percent at the 115th symbol of experience, then remains essentially constant for the next 100 symbols. The 200th prediction is accomplished with a four-

[2] G. H. Hardy and E. M. Wright, *An Introduction to the Theory of Numbers*, Oxford University Press, London, 1962, p. 6.

state machine, the 201st with a three-state machine having but two output symbols of 1. The next two predictions are made by three-state machines having but a single output symbol of 1 each. Prediction, thereafter, is made by a one-state machine with both output symbols being 0, thus reflecting a major "change of policy."[3] As a result, the prediction score once again gradually rises to reach its final value of 81.9 percent after the first 719 symbols of experience. It is evident that this change of policy is entirely reasonable in view of the fact that prime numbers become less and less frequent as the experience grows. These particular results were obtained with a penalty for complexity of 0.01 for each state of the machine, a selection of the best five machines to serve as parents for each next generation, and ten generations being possible prior to each prediction. With different parameters, the trade-off point for the policy change would have shifted. Figure 3.13 reflects the policy in terms of the change in the number of states of the predictor machines.

In order to make the problem have increasing "interest" with greater length of recall, the goal was changed so as to offer increased

[3] It is surprising, and perhaps significant, to note that prior to the change of policy all of the missed primes were second members of twin primes.

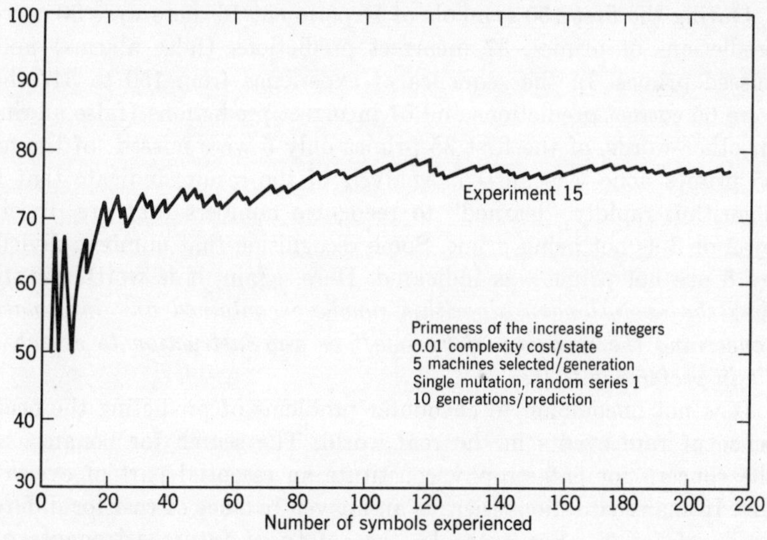

Fig. 3.12. The score for predicting the primeness of each next number in the sequence of positive increasing integers.

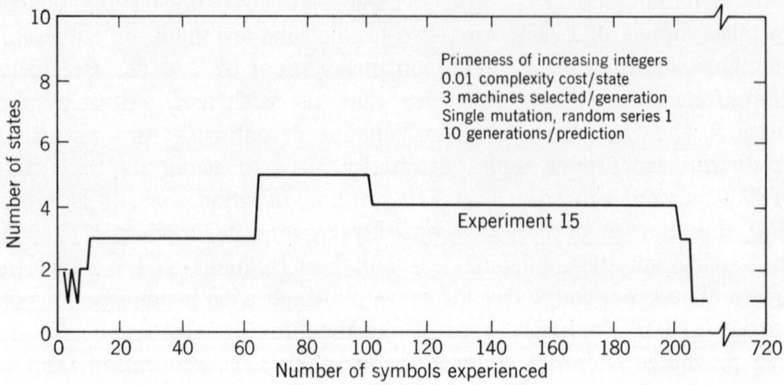

Fig. 3.13. The number of states in the predictor-machines of Experiment 15.

payoff for finding a rare event. Under the new rules, the credit received for correctly predicting a 1 is equal to one plus the number of 0's which preceded that 1; and, similarly, the credit received for properly predicting a 0 is equal to one plus the number of 1's which preceded that 0. This criterion, which places greater value on the prediction of rare events, reduces to the more usual prediction criterion for an equally likely environment.

During the first 150 symbols of Experiment 16 there were 30 correct predictions of primes, 37 incorrect predictions (false alarms) and 5 missed primes. In the sequence of experience from 150 to 547 there were 65 correct predictions and 67 incorrect predictions (false alarms). In other words, of the first 35 primes only 5 were missed; of the next 65 primes none was missed. Analysis of the results indicate that the algorithm rapidly "learned" to recognize numbers that are divisible by 2 or 3 as not being prime. Some recognition that numbers divisible by 5 are not prime was indicated. Here, again, it is worth restating that *the evolutionary algorithm nowhere contained any information concerning the meaning of "prime," or any instruction to search for 1's in preference to 0's.*

It is not uncommon to encounter problems of predicting the occurrence of rare events in the real world. The search for bonanza and the concern for catastrophy constitute an essential part of everyday life. In many situations there is an uneven balance of cost for different kinds of failure. For example, the safety of future astronauts may well depend upon an ability to forecast sudden bursts of intense solar radiation. Certainly, incorrect predictions (false alarms) are costly in that they require spacecraft crewmen to leave their normal assignment and remain within a sheltered area. However, the cost of a missed burst of radiation is far greater; it may constitute disaster for the crew and the mission.

The considerable success achieved in predicting whether or not the next integer is a prime number does not testify against the essential difficulty of the traditional mathematical problem of finding a single formula which would include only the prime numbers. The evolutionary approach does not search for such a formula. In fact, its primary advantage rests upon its allowing a different logical pattern for prediction as each new symbol is considered. Thus, the evolutionary procedure provides a sequence of prediction logics that comprise the best models thus far found within the available expenditure and search capability. In the case of the prime number environment, it should be expected that these best models would tend to be different in view of the nonstationarity. The only necessary consistency among the sequentially best models lies in the fact that they were derived by the same evolutionary processes.

Various means may be employed for reducing the high false alarm rate, and for generally improving the overall efficiency of the prediction process. For example, it might be worthwhile to shorten the length of recall, thereby allowing fewer l's in the sequence against which the predictive fit is measured. Possibly an increase in the probability of adding a state and/or the multiplicity of mutation might increase the efficiency of the search. In view of the logical complexity of the environment, it might be that a greater number of machines should be selected from each generation to serve as possible parents, or that the number of generations prior to each prediction should be changed. An alteration in the length of initial recall may also improve the prospects for a successful prediction of the primeness of the next positive integer. But wait, the purpose of this demonstration was only to indicate the versatility of the evolutionary prediction technique.

As a matter of interest, the evolutionary program was offered portions of the same environmental sequences that were used by Flood in his empirical investigation of stochastic learning.[4] As reported, each of 7 human subjects were required to make 450 consecutive binary choices. After each choice the subject was informed as to the success or failure of his last choice, earning one cent for each success. Each subject repeated the experiment once, but against a

[4] M. M. Flood, "Stochastic Learning Theory Applied to Choice Experiments with Cats, Dogs, and Men," *Behavioral Science,* Vol. 7, July 1962, pp. 289–314.

different "environment." Table 3.2 identifies each trial sequence by the first few symbols (here expressed in octal notation). All scores are taken over the first 200 choices. A comparison of the subject's score to that of the evolutionary program[5] on each trial sequence, reveals a significant difference only in three instances (subject *C*, trial 2; subject *D*, trial 1; and subject *E*, trial 1). In each of these experiments the program was found to be superior.

Table 3.2

A Comparison of the Predictive Ability of the Evolutionary Program to That of Human Subjects—(M. M. Flood)

Initial Sequence	Subject	Trial	Subjects' Score	Evolutionary Program Score
5044014 . . .	*A*	1	0.425	0.460
2755776 . . .	*A*	2	0.445	0.515
0644114 . . .	*B*	1	0.435	0.490
2534756 . . .	*B*	2	0.455	0.435
5744204 . . .	*C*	1	0.445	0.505
2754356 . . .	*C*	2	0.420	0.510
0044304 . . .	*D*	1	0.435	0.560
6534356 . . .	*D*	2	0.545	0.505
0444014 . . .	*E*	1	0.310	0.565
2715756 . . .	*E*	2	0.460	0.450
0144304 . . .	*F*	1	0.495	0.465
2515356 . . .	*F*	2	0.545	0.520
6574356 . . .	*G*	1	0.500	0.495
5444104 . . .	*G*	2	0.570	0.495

In another interesting series of experiments conducted by Wolin, college students were asked to predict each symbol in a sequence of 1000 binary symbols. Each subject was given only one sequence and provided with a visual display of the last 8 symbols and his last 8 predictions at each point in time.[6] The environments were identified as TR-1, TR-3, TR-13, TR-U1A, TR-U3, and WORD. In environment TR-1, odd symbols are equally likely to be *A* or *B* while

[5] Here again the same arbitrary initial machine was used; start state, initial recall of 5 symbols, single mutation, retaining of the best three machines, and five generations per choice.

[6] B. R. Wolin, "Complex Behavior in a 'Single Task' ", *Human Factors,* Vol. 5, pp. 1–5.

all even symbols are *A*. Thus the best expected score for this environment is 0.75. Environment TR-3 is of the same logical structure; however, the even symbols are sure to differ from the preceding symbol. Environment TR-13 is considerably more complex. The odd symbols are generated with a weighting of 0.8 for *A*. The even symbols are always *A*. After each sequence of 10 symbols, the weighting changes to the complement. With this much knowledge, an ideal prediction would approach a score of 0.9. This environment, however, contains additional structure in that the position of the rare symbol within each 10-symbol sequence is moved sequentially through the possible positions; that is, *BAAAAAAAAA*, *AABABABABA*, *AABAAAAAAA*, *BAAABABABA*, and so forth (commas are inserted only as an aid to the reader and were not given to the subjects). TR-13 is therefore deterministic, and could conceivably be perfectly predicted.

Environment TR-U1A is similar in structure to environment TR-1, but consists of word elements. The words *ABA* and *AAAB* are equally likely for odd words, but *AB* always appears as the even word. This is equivalent to an equally likely source of words *ABAAB* and *AAABAB*. Environment TR-U3 is of the same structure as TR-3 but constructed of words. Here *AB* and *AAAB* are equally likely to be odd words, whereas the even words always differ from the preceding odd word. This is equivalent to an equally likely environment composed of the words *ABAAAB* and *AAABAB*. The last environment, WORD, is generated by the recursive structure wherein *AAB* is followed by *AAAA*B 7/10 of the time and by *ABB* 3/10 of the time. *AAAAB* is followed by *AAB* 3/10 of the time and by *ABBB* 7/10 of the time. *ABB* is always followed by *AAAAB*, and *ABBB* is always followed by *AAB*. With complete knowledge of this source, an ideal predictor may expect to be successful 93.4 percent of the time.

Table 3.3 indicates the initial sequence of symbols for each trial (again in octal notation), the designation of that environment, the number of subjects, the ideal score, the score achieved by the same evolutionary program with growth (at each point in time all experienced symbols from that environment are used), the average score of the human subjects, and the score of the same evolutionary program but with recall restricted to the last 10 symbols. Note that in every case these scores appear in rank order. This would appear to indicate that the human operator uses some information from the previous displayed experience. Of course, these results do not reveal the particular mode of human behavior, since these same scores might have been obtained by restricting the recall, weighting the entire recall, or some other procedure.

In order to further validate these results, five additional experiments were carried out against WORD with growth of recall and different random number sequences generating the internal induction. As a result the average score was reduced to 0.827. A rerun against WORD with 10-symbol restricted recall reduced the average score to 0.525. In general, these results are significant only to the extent that they demonstrate the versatility of the evolutionary prediction technique and its favorable score when compared to the specific performance of human subjects.

Table 3.3

A Comparison of the Predictive Ability of the Evolutionary Program to that of Human Subjects (B. R. Wolin)

Initial Sequence	Designation of Environment	Number of Subjects	Ideal Score	Evolutionary Score (with Growth)	Subjects' Average Score	Evolutionary Score (10-Symbol Recall)
521212 . . .	TR-1	6	0.75	0.730	0.690	0.639
523222 . . .	TR-3	6	0.75	0.720	0.695	0.630
4000524 . . .	TR-13	6	1.0	0.730	0.700	0.575
05050524 . . .	TR-U1A	5	0.875	0.865	0.830	0.725
130271 . . .	TR-U3	7	0.92	0.860	0.850	0.725
64125 . . .	WORD	24	0.934	0.845	0.760	0.540

8-SYMBOL PREDICTION EXPERIMENTS

A second program was written to permit the prediction of 8-symbol environments. Unless otherwise indicated, all of the following experiments started with the same arbitrary five-state machine (see Figure 3.14). The recall was permitted to grow, experience starting with 40 symbols before the first prediction. The penalty for complexity was chosen to be 0.01 times the number of states in that machine. Single, double, or triple mutation of each parent machine occurred with equal probability and a maximum of 40 offspring or ten generations were permitted before each successive actual prediction.

The first set of 8-symbol experiments concerned the prediction of an environment composed of a cyclic signal created by repetition of the simple pattern 13576420 that was disturbed by increasing levels

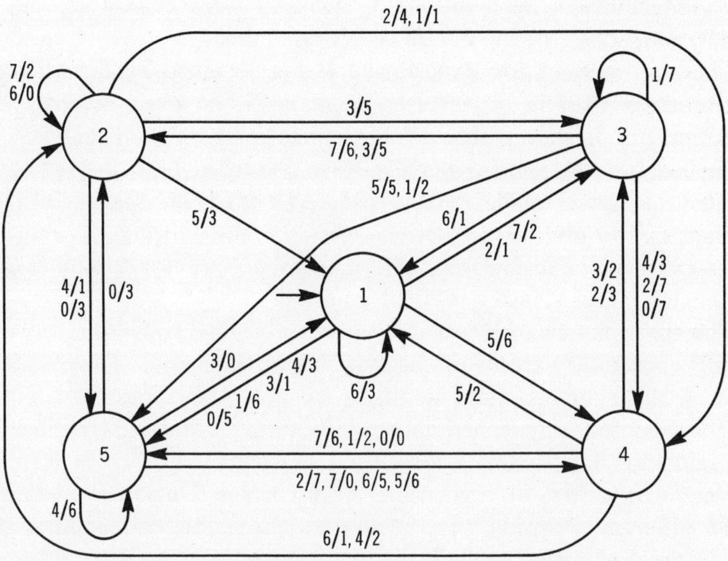

Fig. 3.14. The arbitrary initial 8-symbol machine.

of noise. With the environment consisting only of the undisturbed signal (Experiment 1), the evolutionary technique discovered a perfect one-state predictor-machine within the first eighteen evaluated offspring. The environment for Experiment 2 was generated by corrupting this signal by the equally likely addition of +1 or −1 to certain symbols, these being identified by skipping a number of symbols from the last disturbed symbol in accordance with the next digit drawn from a uniformly distributed random number table. Quite arbitrarily, addition to the symbol 7 and subtraction from the symbol 0 were assumed to leave these symbols undisturbed. Thus, 82.5 percent of the symbols were left undisturbed. As shown in Figure 3.15, 59.3 percent of the first 81 predictions were correct, there being only 6 errors in the last 30 predictions. During the evolution, 3,241 different offspring were evaluated, the predictor-machines growing in size to eight states.

The environment of Experiment 3 was obtained by disturbing the environment used in Experiment 2 once again in the same manner. Thus, 28.9 percent of the symbols were disturbed by ±1, and 1.5 percent were disturbed by ±2, leaving 69.6 percent undisturbed. As shown in Figure 3.15, 39.5 percent of the first 81 predictions were correct, there being a general increase in score in the last 20 predictions. During the evolution, 3,236 different offspring were evaluated, the predictor-machines growing in size to fifteen states.

The environment for Experiment 4 was obtained by disturbing the environment used in Experiment 3, again in the same manner. Thus, 37.0 percent of the symbols were disturbed by ±1, 2.5 percent were disturbed by ±2, and 0.1 percent were disturbed by ±3, leaving 60.4 percent of the symbols undisturbed. As shown on Figure 3.15, 23.5 percent of the first 81 predictions were correct. During the evolution, 3,214 different offspring were evaluated, the predictor-machines growing in size rather steadily to nineteen states.

The environment of Experiment 5 was obtained by disturbing a randomly chosen 50 percent of the symbols in the signal. Thus, 43.8 percent of the symbols were disturbed by ±1 (the difference being due to the adopted rule concerning addition to 7 and subtraction from 0), leaving 56.2 percent undisturbed. As shown in Figure 3.15, 22.2 percent of the first 81 predictions were correct. During the evolution, 3,195 different offspring were evaluated, the predictor-machines growing in size somewhat irregularly to seventeen states. It would appear that this last increase in the noise level (from 39.67 percent in Experiment 4 to 43.8 percent in Experiment 5) resulted in significantly degraded prediction of the environment.

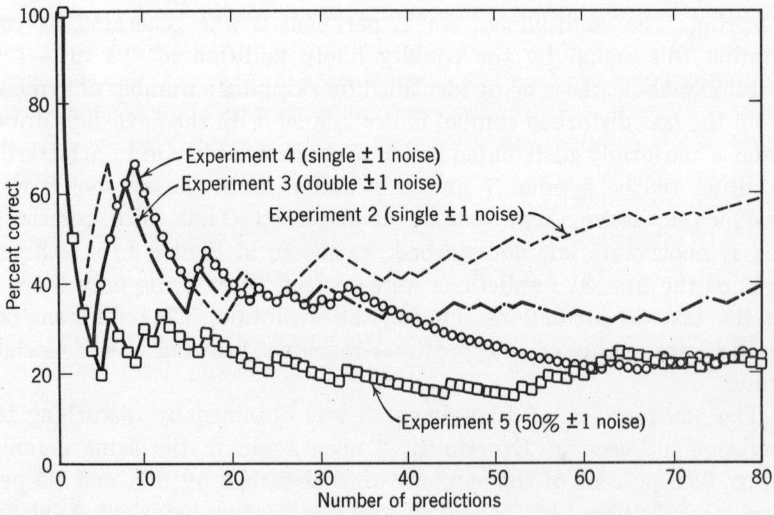

Fig. 3.15. Evolutionary prediction of the 8-symbol disturbed cyclic environment.

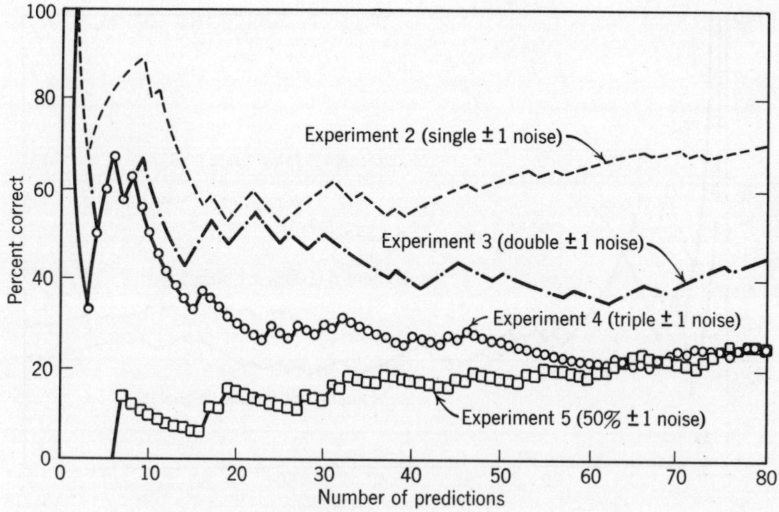

Fig. 3.16. Evolutionary prediction of the cyclic signal.

Figure 3.16 indicates the degree of correspondence between the sequence of predictions and the signal in these experiments. Note that after the first 76 predictions the signal was predicted in Experiment 5 as well as it was in Experiment 4, in spite of the fact that a larger percentage of the signal symbols had been disturbed. This may be due to the fact that in the last experiment the symbols remained closer to the original signal. Such "consideration" for the magnitude of deviations is a result of using a distance-weighted error matrix, the weighting in this case being the magnitude of the symbol difference. In essence, this choice converts the nominal scale of symbols to an ordinal scale.

It is of interest to examine each of the predictor-machines as representations of the periodic properties of the environment. The *characteristic cycle* for any finite-state machine is found by starting it in its initial state together with the first symbol of the recall and then driving it by each of its successive output symbols until the output sequence is periodic. All of the characteristic cycles in Experiment 2 were 8 symbols in length. The first 73 predictions corresponded perfectly with the pattern of the signal, but this "insight" was lost in the later predictions which were in error by 1 or 2 symbols. After the 14th prediction the characteristic cycle remained 13576430.

The higher noise level of the third experiment resulted in character-

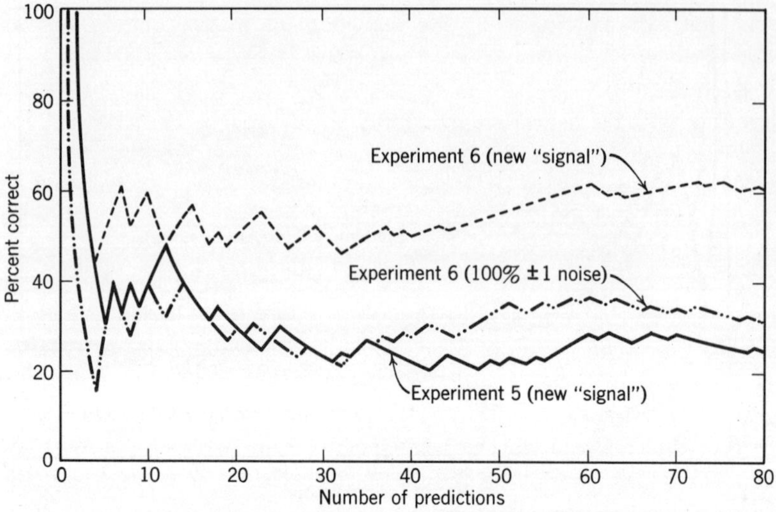

Fig. 3.17. Comparison of experiments 5 and 6.

istic cycles of varying length until the 26th prediction. From then until the 70th prediction, the characteristic cycle remained 1357643113576430, this being in error 1 symbol out of every 2 cycles of the signal. As expected, the result of Experiment 4 was more erratic, with the length of characteristic cycle jumping from 8 to 16 symbols and remaining the same after the 73rd prediction. Prediction of each of the last 24 characteristic cycles was 62.5 percent correct. Experiment 5 revealed even greater variability in the characteristic cycles. A majority of these were 8 or 16 symbols in length and reflected the basic pattern of the signal although there was little one-to-one correspondence.

The environment for Experiment 6 was generated by disturbing every symbol of the signal by +1 or −1 with equal probability. At first glance it is surprising to find the prediction of the signal improved as shown in Figure 3.17, but note that with the disturbance of every symbol one aspect of the randomness of the environment was removed. In essence, the signal had taken on a new form, the boundaries of the original signal 24677531 or 02465310 each having equal probability at each point in time. In the first 81 predictions this new signal was properly identified 70.4 percent of the time. In fact, the characteristic cycle of the last predictor-machine was 0226773110266533l. This can be seen to lie on the boundaries of the original signal except for 1 symbol of every 17. In order to provide a basis for comparison, the percent correct prediction of the environment is also shown, as well as the result of Experiment 5 with respect to the new signal.

It is natural to inquire as to the extent to which the prediction capability will be degraded by "wild noise" (each disturbed symbol being replaced by a randomly chosen symbol from the input alphabet). The environments of Experiments 7 and 8 were generated by imposing this kind of disturbance on the original signal once and twice, respectively. As expected, the ability of the evolutionary program to predict the environment, as shown in Figure 3.18, was somewhat poorer than in the comparable Experiments 2 and 3. Figure 3.19 indicates the degree of correspondence between the sequence of predictions and the signal in these same experiments. Here, again, the additional degree of randomness within the noise degrades the performance. Carrying this noise to the extreme results in a perfectly random environment which cannot be predicted. Experiment 9 revealed no significant ability of the evolutionary program in its attempt to predict such an environment.

The introduction of randomness always introduces questions of repeatability. In order to examine this point, Experiment 2 was repeated nine additional times, the results being shown in Figure 3.20. As expected, the variability is an inverse function of the score.

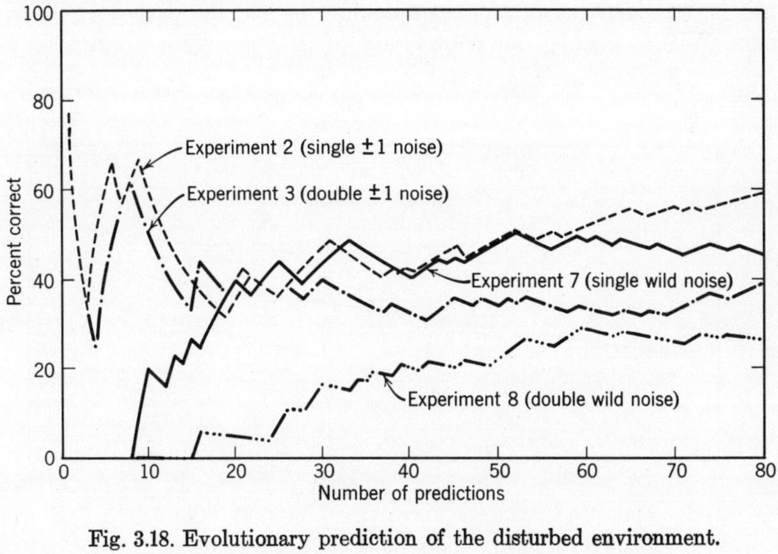

Fig. 3.18. Evolutionary prediction of the disturbed environment.

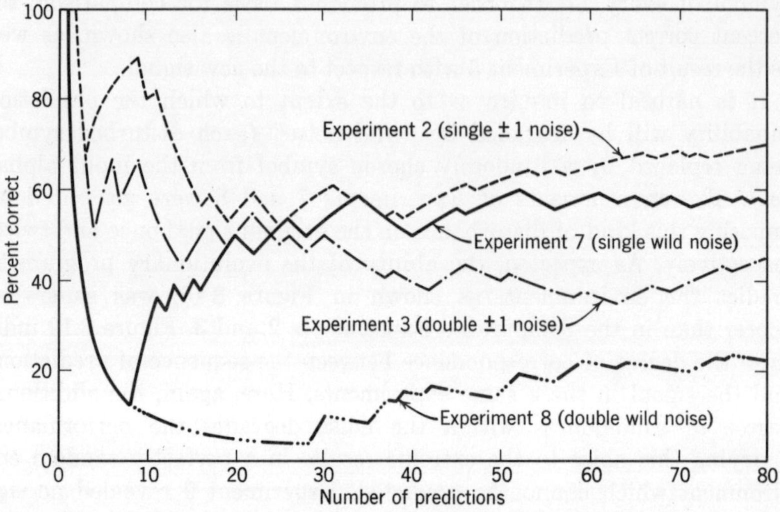

Fig. 3.19. Evolutionary prediction of the signal.

The second set of experiments was concerned with the prediction of purely stochastic environments. Experiment 10 required the prediction of a zeroth-order 8-symbol Markov environment generated by combining two normally distributed variables over 5 symbols each to yield the bimodal environment shown in Table 3.4. A sample of 269 symbols was generated, this sample having the relative frequencies shown. With a penalty for complexity of 0.01 per state the predictor-

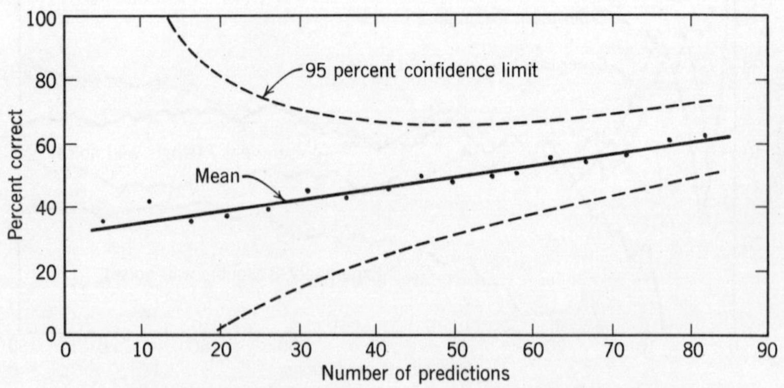

Fig. 3.20. Statistics of ten repetitions of experiment 2.

machines grew in size to thirty-one states. The relative frequency of each symbol in the sequence of predictions suitably reflects the relative frequencies shown in the sample. Experiment 11 was conducted with an increase of the penalty for complexity in order to decrease the sensitivity of the sequence of predictions to the difference in the relative frequency of the modes. With a penalty for complexity of 0.1 per state the predictor-machines grew in size only to eleven states and, as shown in Table 3.4, the predictions were in closer correspondence with the statistics of the environment.

Table 3.4

Symbol	Intended Probability	Relative Frequency of Sample	Experiment 10. Relative Frequency of Predictions	Experiment 11. Relative Frequency of Predictions
0	0.0116	0.00374	0.08	0.005
1	0.06785	0.0825	0.115	0.061
2	0.3413	0.299	0.223	0.303
3	0.07925	0.1050	0.058	0.081
4	0.07925	0.0787	0.017	0.015
5	0.3413	0.348	0.521	0.520
6	0.06785	0.0637	0.033	0.015
7	0.0116	0.0187	0.025	0

Experiment 12 required the prediction of a zeroth-order 4-symbol Markov environment, the arbitrarily chosen probabilities being 0.1, 0.2, 0.3, and 0.4. This information, as prior knowledge, would dictate the continual prediction of the most probable symbol giving the asymptotic score of 40 percent. At the other extreme, perfectly random prediction would have an expected score of 25 percent. As shown in Figure 3.21 the evolutionary score settled between these extremes, thus demonstrating the purposeful extraction of information from the previous symbols.

The first-order environment of Experiment 13 was generated in such a way as to produce theoretically the results shown in Table 3.5. Specifically, after an initial arbitrary symbol, each next symbol of the environment was generated utilizing the relative frequency distributions of the next symbol, given a preceding symbol as indicated by the rows of the matrix shown in Table 3.5. Here the symbols in the top line are the next symbols following the preceding symbols in the

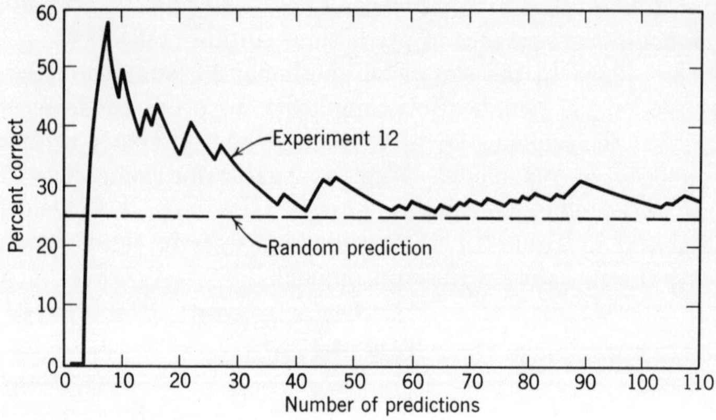

Fig. 3.21. Evolutionary prediction of zeroth-order Markov environment.

left column. The actual environment generated had the transition matrix of relative frequencies shown in Table 3.6. The marginal frequencies of this environment were 0.236, 0.241, 0.249, and 0.274, respectively.

Table 3.5

	0	1	2	3
0	0	0.8	0.1	0.1
1	0.1	0	0	0.9
2	0.9	0.1	0	0
3	0	0.1	0.8	0.1

Table 3.6

	0	1	2	3
0	0	0.822	0.071	0.107
1	0.035	0	0	0.965
2	0.915	0.085	0	0
3	0	0.077	0.862	0.061

With only the knowledge that the process is first-order it was possible to attain the score of 89.5 percent on the 200th prediction in the manner shown in Figure 3.22. *But even without this knowledge the evolutionary prediction technique attained this same score.* Analysis of the sequence of predictions revealed that at the end of this experiment the environment was properly characterized by the maximum transition probabilities of each row. Other experiments were conducted on first- and second-order processes with satisfactory results.

To generate a more difficult environment, the powers of 2 and 3 were rank ordered and reduced modulo 8, thus producing the sequence 1234010300103001003010030100300 . . . . This environment was predicted in Experiment 14. Incidentally, the problem of rank ordering the powers of 2 and of 3 may be simplified by writing all of the powers in terms of a common base. For example, the powers of 3 may be written in terms of powers of 2 taking $2^{1.584925}$ as an approximation. Since $2^x > 2^v$ if, and only if, $x > v$ the problem is reduced to one of ordering the positive integers and the integral multiples of 1.584925. Since this basic exponent is in error by less than 10^{-7}, 10^7 powers of 3 may be considered before an incorrect ordering could occur.

Taking a modulus of this increasing series constrains the numbers to the alphabet available for the evolutionary program. With respect to modulo 8 the powers of 2 yield the residues 2, 4, followed by 0's, whereas the powers of 3 are alternately 3 and 1. To see this note that $2^1 = 2 \pmod 8$, $2^2 = 4 \pmod 8$ and $2^3 = 0 \pmod 8$. For $k > 3$,

$$2^k = 2^3 \cdot 2^{k-3} = 0 \cdot 2^{k-3} \pmod 8 = 0 \pmod 8$$

Similarly, $3^1 = 3 \pmod 8$ and $3^2 = 1 \pmod 8$ so that for n even (that is, $n = 2k$),

$$3^{2k} = (3^2)^k = 1^k \pmod 8 = 1 \pmod 8$$

while for n odd (that is, $n = 2k + 1$),

$$3^{2k+1} = 3^{2k} \cdot 3 = 1 \cdot 3 \pmod 8 = 3 \pmod 8$$

With respect to modulo 7, the powers of 2 form the repeating sequence 2, 4, 1, 2, 4, 1, . . . while powers of 3 form the repeating se-

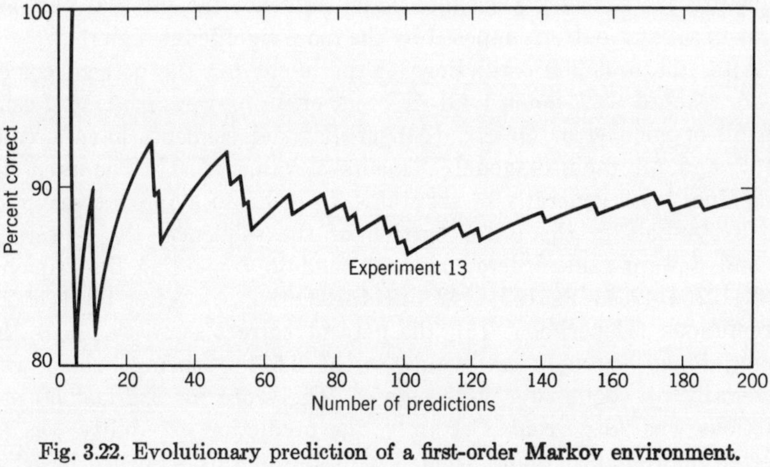

Fig. 3.22. Evolutionary prediction of a first-order Markov environment.

quence 3, 2, 6, 4, 5, 1, 3, 2, 6, 4, 5, 1, ... This can be seen by noting that $2^3 = 1 \pmod 7$, $2^{3k} = (2^3)^k = 1^k \pmod 7$, and $1^k = 1 \pmod 7$;

$$2^{3k+1} = (2^{3k}) \cdot 2 = 1 \cdot 2 \pmod 7 = 2 \pmod 7$$

while

$$2^{3k+2} = (2^{3k})\, 2^2 = 1 \cdot 4 \pmod 7 = 4 \pmod 7$$

Similarly, since $3^6 = 1 \pmod 7$,

$$3^{6k} = (3^6)^k = 1^k \pmod 7 = 1 \pmod 7$$

$$3^{6k+1} = 3^{6k} \cdot 3 = 3 \pmod 7$$

$$3^{6k+2} = 3^{6k} \cdot 3^2 = 3^2 \pmod 7 = 9 \pmod 7 \quad \text{and } 9 = 2 \pmod 7$$

$$3^{6k+3} = 3^{6k} \cdot 3^3 = 27 \pmod 7 \quad \text{and } 27 = 6 \pmod 7$$

$$3^{6k+4} = 3^{6k} \cdot 3^4 = 81 \pmod 7 \quad \text{and } 81 = 4 \pmod 7$$

$$3^{6k+5} = 3^{6k} \cdot 3^5 = 243 \pmod 7 \quad \text{and } 243 = 5 \pmod 7$$

Therefore, to form the sequence of residues modulo 8, it is only necessary to order the integers and the integral multiples of 1.584925; that is, the sequence 2, 4, followed by all 0's is inserted in successive positions corresponding to the integers while the repetitive sequence 3, 1, . . . is inserted in the successive positions corresponding to the multiples of 1.584925. Similarly, for modulo 7 the integer sequence 2, 4, 1, 2, 4, 1, . . . is interlaced with the sequence 3, 2, 6, 4, 5, 1, 3, . . . which corresponds to the multiples of 1.584925. The resulting sequences arc not periodic although subperiods are embedded throughout. This lack of periodicity results since in higher multiples of 1.584925 the successive decimal digits influence the integral part and thus upset any pattern imposed by the more significant digits.

After the first 300 predictions (Experiment 14) the percent correct score reached 88.7. Some 1,401 different offspring were evaluated using an all-or-none error matrix. (All off-diagonal elements have a value of 7 and all main-diagonal elements a value of 0.) The predictor machines were generally of six states. In order to avoid the reduction to 3 symbols in the latter portion of the sequence, the powers of 2 and 3 were rank ordered reduced modulo 7 yielding the sequence 123412264142541124312246142451214312246l . . . . After the first 216 predictions (Experiment 15), the percent correct score was 56.6, this being found through the evaluation of 3,671 offspring, which were generally of about twenty-one states. The score for the last 50 predictions was 78 percent. Certainly, the prediction capability was far better than chance would yield. Prediction of this sequence based on the most probable symbol up to each point in time yields a score of only 8.5 percent.

Experiments were conducted in order to evaluate the evolutionary technique as a means for detecting the existence of correlation between variables, this correlation to be used to enhance the sequential prediction of one or more of the sensed variables. In Experiment 16, a random sequence of binary symbols was presented to an arbitrarily chosen finite-state machine, *M* (shown in Table 3.7) starting in state 1. Note

Table 3.7
Machine *M*

Present State	1	1	2	2	3	3	4	4	5	5
Input Symbol	0	1	0	1	0	1	0	1	0	1
Next State	3	5	1	2	1	4	5	3	2	1
Output Symbol	2	0	2	3	1	1	3	0	0	2

that four different symbols comprise the output alphabet of this machine. The evolutionary program was required to predict (based solely on the previously observed output symbols) each next symbol in the output sequence, this being predictable only to the extent that the structure of the machine *M* is superimposed upon the otherwise random driving signal. The percent correct score, shown in Figure 3.23,

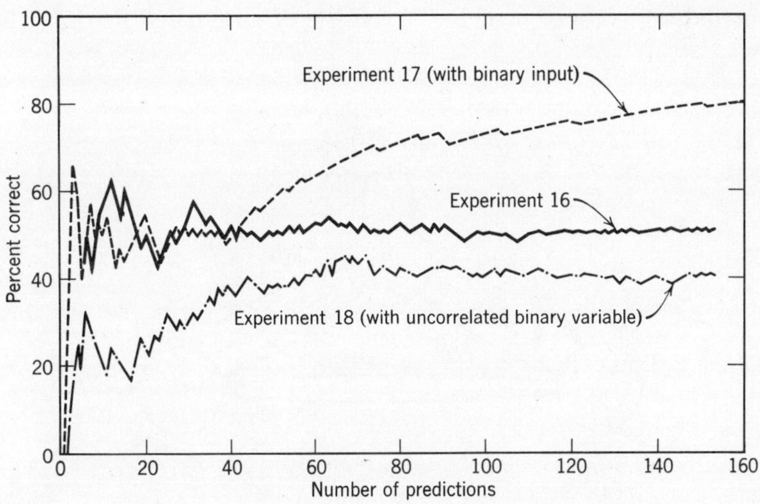

Fig. 3.23. Evolutionary prediction of the 4-symbol environment from machine *M*.

reveals increasing stability as the score settles around 51 percent after the first 152 predictions (with 20 symbols as the initial recall).

This prediction score should be considerably improved if the program is permitted to evolve finite-state representations on the basis of both the 4-symbol output sequence and the 2-symbol input sequence. In fact, with this additional information, the prediction score should asymptotically approach 100 percent as the recall increases, in view of the deterministic nature of machine M. Experiment 17 was conducted to investigate whether or not this would take place through the evolutionary technique. At each point in time, two-input single-output finite-state machines were evaluated in terms of their predictive fit of the known recall of the 4-symbol output sequence while driven by the preceding recall of both the 4-symbol output and 2-symbol input sequences. The percent correct score, shown in Figure 3.23, attains a value of 80 percent.

In some sense, the sequence of evolved double-input single-output machines should provide an increasingly accurate representation of the transduction performed by machine M. Figure 3.24 is a graphical representation for the finite-state machine that evolved after 160 predictions. This machine, N, has the states designated by letters in order to avoid any confusion with the number-designated states of machine M. It is possible to compare machine N to machine M in the following way: Examine each state of M in order to determine if there is some state in machine N which, in some sense, will perform the same function when given the same sequence of input signals. (Note that the input to each state of machine N consists of an ordered pair of sym-

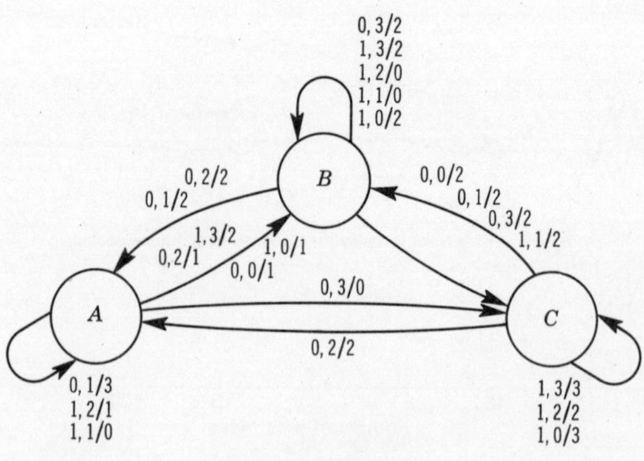

Fig. 3.24. Machine N.

bols, the first being the input to the machine *M*, and the second being the previous output of machine *M* before it entered that state.)

To illustrate, consider state 1 of machine *M*. Since the outputs of the transitions that enter this state are either 1 or 2, the possible corresponding inputs to machine *N* are (0, 1), (0, 2), and (1, 1) and (1, 2). Examination of state *B* of machine *N* shows that 0, 1 and 0, 2 both yield the output symbol 2, which corresponds to the output of state 1 when it receives an input symbol of 0. Similarly 1, 1, and 1, 2 produce an output of 0 which corresponds to the output of state 1 when it receives input symbol 1. Thus, state 1 of machine *M* may be said to be contained in state *B* of machine *N* in the sense that machine *N* will yield the same response if given a corresponding sequence of inputs. Following this logic, it can be shown that state 3 is contained in state *A* and state 5 is contained in state *B*.

Viewing the problem in the large, it is often essential to formulate hypotheses concerning the logical relation between two or more separately sensed variables. It would appear that the evolutionary technique offers a means for finding multiple-input predictor-machines which can be translated into reasonable representations for the observed logical relationship.

Experiment 18 was conducted to confirm the claim that the evolutionary technique had, indeed, extracted useful information from the correlative input variable in Experiment 17. In Experiment 18 two-input single-output machines were evolved in the very same manner as before, but in this case the binary variable that was offered was uncorrelated with the output sequence taken from *M* in that it was independently generated. The percent correct score, shown in Figure 3.23 remains below the corresponding scores that resulted from the previous two experiments. In other words, the additional information that was furnished in the form of an independent binary variable formed a "distraction" resulting in poorer predictions.

In view of the apparent need, a formal technique was devised that permits expression of the set of finite-state machines functionally contained within any given machine. Thus it becomes possible to translate each predictor-machine into a set of hypotheses concerning the transduction logic of the environment. It is reasonable to expect that the higher the score of the predictor-machine, the greater will be the correspondence of the hypothesis with the logical properties of the environment.

To make this presentation complete and self-contained some background material is introduced concerning finite-state machines and their formal representation.

For a set A, the free semigroup[6] on A, written ΣA, is the set of all finite sequences of elements from A with the operation being that of concatenation of sequences; that is, if $a = a_1 \cdots a_n$ and $b = b_1 \cdots b_m$ are two sequences, then ab is the sequence $a_1 \cdots a_n b_1 \cdots b_m$. The length, $l(a)$, of a sequence is the number of elements in the sequence, so that $l(a) = n$ and $l(ab) = n + m$. If the null sequence, denoted by ϕ, is adjoined to ΣA with the rule $a\phi = \phi a$ for all a in ΣA, then the resulting semigroup is denoted by $\overline{\Sigma A}$ and is termed the free semigroup on A with identity. The proof that concatenation of sequences is associative, that is, $(ab)c = a(bc)$, is immediate, and hence both ΣA and $\overline{\Sigma A}$ are semigroups.

With each element, a, of a semigroup S there is associated the left translation function, L_a, from S into itself defined by $L_a(x) = ax$ for all x in S. Note that if S has an identity, ϕ, then $L_\phi(x) = x$. Further, if f is any function from a semigroup S into a set B, then the composition function of f and L_a defined by $(f \circ L_a)(x) = f(L_a(x)) = f(ax)$ is also a function from S into B for each a in S. In particular, if S has an identity, ϕ, then $f \circ L_\phi = f$.

A finite-state machine is a quintuple $(A, B, Q, \delta, \lambda)$ where A is a finite set called the "input alphabet," Q is a finite set called the "states," λ is the "next-state function" which is a mapping of $A \times Q$ into Q, and δ is the "output function" which is a mapping of $A \times Q$ into the finite set B. At times, a particular state q_0 (a member of the set Q) is also included in this definition as the "initial state," the intent being that all inputs to the machine should begin with the machine in the state q_0.

The function λ and δ can be extended to the functions $\bar{\lambda}$ and $\bar{\delta}$, respectively, on $\Sigma A \times Q$ according to the following recursive definitions. If $l(a) = 1$, so that a belongs to A, then $\bar{\lambda}(a, q) = \lambda(a, q)$ for all q. If $l(a) = n > 1$, so that $a = a_1 \cdots a_{n-1}a_n$, then $\bar{\lambda}(a_1 \cdots a_{n-1}a_n, q) = \lambda(a_n, \bar{\lambda}(a_1 \cdots a_{n-1}, q))$. Similarly, if $l(a) = 1$ so that a belongs to A, then $\bar{\delta}(a, q) = \delta(a, q)$ for all q. If $l(a) = n > 1$, so that $a = a_1 \cdots a_{n-1}a_n$, then $\bar{\delta}(a_1 \cdots a_{n-1}, q) = \delta(a_n, \bar{\lambda}(a_1 \cdots a_{n-1}, q))$. Intuitively, $\bar{\lambda}(a, q)$ is the final state of the machine if the sequence, a, is introduced as the input with the machine in state q, while $\bar{\delta}(a, q)$ is the last output symbol. By induction on the length of b, it can be shown that for all a and b belonging to ΣA and q belonging to Q,

$$\bar{\lambda}(ab, q) = \bar{\lambda}(b, \bar{\lambda}(a, q))$$

[6] A semigroup is a nonempty set of elements, Z, together with an associative binary operator $\circ$ on Z. That is, for all a, b, and c in the set Z, $a \circ b$ is in Z and $(a \circ b) \circ c = a \circ (b \circ c)$.

and

$$\tilde{\delta}(ab, q) = \tilde{\delta}(b, \tilde{\lambda}(a, q))$$

For each state q there is a function f_q, mapping the input set ΣA into the output set B, which can be defined by $f_q(a) = \tilde{\delta}(a, q)$ for all a in ΣA. There is a set of functions, $[f_q]$, for any machine, each member of this set mapping ΣA into B. Further, consider the composition mapping $f_q \circ L_a$ for any state q and any input sequence a. Clearly, for all x in ΣA, $f_q \circ L_a(x) = f_q(ax) = \tilde{\delta}(x, \tilde{\lambda}(a, q)) = f_{\tilde{\lambda}(a,q)}(x)$. Therefore $f_q \circ L_a = f_{q'}$ where $q' = \tilde{\lambda}(a, q)$; that is, the set of functions $[f_q]$ is closed under composition with the set of left translations, $[L_a]$, on ΣA. Note that there may be fewer functions than states, since different states may give rise to the same function. If there are exactly as many functions as states, then the machine is said to be reduced.

Conversely, suppose that ΣA is the free semigroup on a finite set A and $[f_q \mid q \in Q]$ is a finite set of distinct functions mapping ΣA into a finite set B. Further, let the set $[f_q]$ be closed under composition with the set of left translations, $[L_a]$, on ΣA; that is, for any q in Q and any sequence a, $f_q \circ L_a = f_{q'}$ for some q' in the set Q. Then there is a finite-state machine with input alphabet A, output alphabet B, and one state for each element q in Q such that $[f_q]$ is exactly the set of functions associated with the machine as above.

Now return to the experiment in which a machine N of the form $N = (A \times B, B, P, \lambda_N, \delta_N, p_0)$ was developed to simulate a machine M of the form $M = (A, B, Q, \lambda_M, \delta_M, q_0)$, and the problem of reconstructing M from N. Before defining the concept of one machine being embedded in another, it is convenient to introduce some additional concepts.

If M is the machine $(A, B, Q, \lambda_M, \delta_M, q_0)$ then B_q is the subset of B consisting of those b's for which there is a q_1 in Q and an a in A for which $q = \lambda_M(a, q_1)$ and $b = \delta_M(a, q_1)$.

If N is the machine $(A \times B, B, P, \lambda_N, \delta_N, p_0)$ then the sequence $(a_0, b_0) \cdots (a_n, b_n)$ is an allowable sequence for the machine N if $\tilde{\delta}_N((a_0, b_0) \cdots (a_i, b_i), p_0) = b_{i+1}$ for all $i < n$.

Lemma 1

For N as above, for any b_0 in B and for any sequence $a_0 \cdots a_n$ in ΣA there is a unique sequence $b_1 \cdots b_n$ such that $(a_0, b_0) \cdots (a_n, b_n)$ is an allowable sequence for N.

Let $b_1 = \delta_N((a_0, b_0), p_0)$ and by induction, define $b_i = \tilde{\delta}_N((a_0, b_0) \cdots (a_{i-1}, b_{i-1}), p_0)$. Then $(a_0, b_0) \cdots (a_n, b_n)$ is an allowable sequence for N. If $b_1' \cdots b_n'$ is another such sequence, then $b_1 = \delta_N((a_0,$

$b_0), p_0) = b_1'$. If $b_j = b_j'$ for $j < i$, then since $b_i = \bar{\delta}_N((a_0, b_0) \cdots (a_{i-1}, b_{i-1}), p_0) = b_i'$, by induction $b_i = b_i'$ for $1 \leq i \leq n$.

For machines M and N as above,[7] q in Q is embedded in p belonging to P relative to b in B if for all a in A, $\delta_N((a, b), p) = \delta_M(a, q)$. M is embedded in N if (1) for some b_0 in B, q_0 is embedded in p_0 relative to b_0; and (2) for $q = \tilde{\lambda}_M(a_0 \cdots a_n, q_0)$, if $(a_0, b_0) \cdots (a_n, b_n)$ is the allowable sequence associated with $a_0 \cdots a_n$ and b_0 by Lemma 1, then for all i, $1 \leq i \leq n$, $q_i = \tilde{\lambda}_M(a_0 \cdots a_{i-1}, q_0)$ is embedded in $p_i = \tilde{\lambda}_N((a_0, b_0) \cdots (a_{i-1}, b_{i-1}), p_0)$ relative to b_i.

q in Q is strongly embedded in p belonging to P if for all a in A and each b in B_q, $\delta_M(a, q) = \delta_N((a, b), p)$. (Note that if q is strongly embedded in p, it is embedded in p relative to any b in B_q.) M is strongly embedded in N if (1) for some b_0 in B, q_0 is embedded in p_0 relative to b_0; and (2) for $q = \tilde{\lambda}_M(a_0 \cdots a_n, q_0)$, if $(a_0, b_0) \cdots (a_n, b_n)$ is the allowable sequence associated with $a_0 \cdots a_n$ and b_0 by Lemma 1, then for all i, $1 \leq i \leq n$, $\tilde{\lambda}_M(a_0 \cdots a_{i-1}, q_0)$ is strongly embedded in $p_i = \tilde{\lambda}_N((a_0, b_0) \cdots (a_{i-1}, b_{i-1}), p_0)$.

Theorem I

If M is strongly embedded in N, then it is embedded in N.

Proof:

For $q = q_0$, the definitions are identical. For $q = \tilde{\lambda}_M(a_0 \cdots a_n, q_0)$, in the definition of embedding, b_n belongs to B_q. As noted in the definition of strong embedding, q is embedded relative to any b in B_q and so, in particular, relative to b_n.

Lemma 2

Let M be embedded in N with q_0 embedded in p_0 relative to b_0. For any sequence $a_0 \cdots a_n$ in ΣA, if $(a_0, b_0) \cdots (a_n, b_n)$ is the unique allowable sequence for N of Lemma 1, then for $0 \leq i \leq n$, $\bar{\delta}_M(a_0 \cdots a_i, q_0) = \bar{\delta}_N((a_0, b_0) \cdots (a_i, b_j), p_0)$.

Proof:

By Lemma 1, the sequence $b_1 \cdots b_n$ must be the sequence as given in the definition of embedded. By this definition, each q_i is embedded in p_i relative to the corresponding b_i and hence $\bar{\delta}_M(a_0 \cdots a_i, q_0) = \bar{\delta}_N((a_0, b_0) \cdots (a_i, b_i), p_0)$ for $0 \leq i \leq n$.

[7] In order to preclude pathological cases and to make the statements of theorems simpler, it is assumed Q is restricted to those states q such that $q = q_0$ or $q = \tilde{\lambda}_M(a, q_0)$ for some a in ΣA. Similarly for P. In other words, states which cannot be reached are deleted from consideration. There is no loss in the functional aspects of the machines.

Theorem II

If M is embedded in N, then the reduced machine of M is embedded in N.

This theorem can be derived directly from Theorems III and IV, therefore the proof will be given after the proofs of these theorems.

In the experiment being considered, the problem was: Given the machine N, is there a machine M, or possibly machines strongly embedded or embedded in N? If so, how does one find such a machine or machines? Both of these questions are answered by the following procedure which constructs such a machine.

Let b_0 be any element in B, then define the function f on ΣA to B as follows:[8] For any sequence $a_0 \cdots a_n$ in ΣA, $f(a_0 \cdots a_n) = \bar{\delta}_N((a_0, b_0) \cdots (a_n, b_n), p_0)$ where $(a_0, b_0) \cdots (a_n, b_n)$ is the unique allowable sequence associated by Lemma 1 with b_0 and the sequence $a_0 \cdots a_n$. The following Lemma will set the stage for the desired construction.

Lemma 3

The set of functions $\{f \circ L_a \mid a \text{ in } \Sigma A\}$ is finite.

Proof:

Consider the function $f \circ L_{a_0 \cdots a_n}$. Let $a_0' \cdots a_k'$ be any sequence in ΣA and $(a_0, b_0) \cdots (a_n b_n)(a_0', b_0') \cdots (a_k', b_k')$ be the allowable sequence associated with $a_0 \cdots a_n a_0' \cdots a_k'$. Then

$$\begin{aligned} f \circ L_{a_0 \cdots a_n}(a_0' \cdots a_k') &= f(a_0 \cdots a_n a_0' \cdots a_k') \\ &= \bar{\delta}_N((a_0, b_0) \cdots (a_n, b_n)(a_0', b_0') \cdots (a_k', b_k'), p_0)) \\ &= \bar{\delta}_N((a_0', b_0') \cdots (a_k', b_k'), \bar{\lambda}_N((a_0, b_0) \cdots (a_n, b_n), p_0)) \\ &= \bar{\delta}_N((a_0', b_0') \cdots (a_k', b_k'), p_{n+1}) \end{aligned}$$

where

$$p_{n+1} = \bar{\lambda}_N((a_0, b_0) \cdots (a_n, b_n), p_0) \quad \text{and} \quad b_0' = \bar{\delta}_N((a_0, b_0) \cdots (a_n, b_n), p_0).$$

By Lemma 1 for $0 < i \leq k$,

$$\begin{aligned} b_i &= \bar{\delta}_N((a_0, b_0) \cdots (a_n, b_n)(a_0', b_0') \cdots (a'_{i-1}, b'_{i-1}), p_0) \\ &= \bar{\delta}_N((a_0', b_0') \cdots (a'_{i-1}, b'_{i-1}), p_{n+1}) \end{aligned}$$

[8] f should be denoted by f_{b_0}, since in general a different choice of b_0 may give rise to a different function. Wherever the context requires the subscript for clarity, it will be included.

Hence, $(a_0', b_0') \cdots (a_k', b_k')$ is the unique allowable sequence for the machine $N_{n+1} = (A \times B, B, P, \lambda_N, \delta_N, p_{n+1})$ associated with b_0' and $a_0' \cdots a_k'$. The function $f \circ La_0 \cdots a_n$ is thus completely determined by the pair b_0' and p_{n+1}. But B and P are finite sets, so there are only a finite number of pairs (b_0', p_{n+1}); that is, there are only a finite number of distinct functions in the set $\{f \circ L_a | a \text{ in } \Sigma A\}$.

Suppose the set of functions consist of $f, f \circ L_{a_1}, \cdots, f \circ L_{a_k}$. Let $Q = \{q_0, \cdots, q_k\}$ and let q_0 correspond to f and q_i correspond to $f \circ L_{a_i}$ for $0 < i \leq k$. Then $M' = (A, B, Q, \lambda_M', \delta_M', q_0)$ is a finite state machine with $\bar{\lambda}_M'(a, q_i) = q_j$ where $f \circ L_{a_i a} = f \circ L_{a_j}$, unless $f \circ L_{a_i a} = f$ in which case $j = 0$, and with $\bar{\delta}_M'(a, q_i) = f(a_i a)$. Further, $f, f \circ L_{a_i}, \cdots, f \circ L_{a_k}$ are the functions on ΣA to B which are associated with the states of M', and M' is a reduced machine.

Theorem III

M' as defined above is embedded in N.

Proof:

Let q_0 be the initial state of M' corresponding to the function f, then for any a in A;

$$\delta_M'(a, q_0) = f(a) \text{ by definition of } M'$$

$$= \delta_N((a, b_0), p_0) \text{ by definition of } f$$

Hence, q_0 is embedded in p_0 relative to b_0.

Let $a_0 \cdots a_n$ be any sequence in ΣA and let $(a_0, b_0) \cdots (a_n, b_n)$ be the allowable sequence associated with it so that for $1 \leq i \leq n$, $b_i = \bar{\delta}_N((a_0, b_0) \cdots (a_{i-1}, b_{i-1}) p_0)$. If $q_i = \bar{\lambda}_M'(a_0 \cdots a_{i-1}, q_0)$ and $p_i = \bar{\lambda}_N((a_{i-1}, b_{i-1}), p_0)$, then it is required to show q_i is embedded in p_i relative to b_i for $1 \leq i \leq n$. Let k be any integer such that $1 \leq k \leq n$ and let a be any element in A. It is required to show $\delta_M'(a, q_k) = \delta_N((a, b_k), p_k)$.

$$\text{Now, } \delta_M'(a, q_k) = \delta_M'(a, \bar{\lambda}_M'(a_0 \cdots a_{k-1}, q_0)) \text{ by definition of } q_k$$

$$= \bar{\delta}_M'(a_0 \cdots a_{k-1}a, q_0)$$

$$= f(a_0 \cdots a_{k-1}a) \qquad \text{by definition of } M'$$

$$= \bar{\delta}_N((a_0, b_0) \cdots (a_{k-1}, b_{k-1})(a, b_k), p_0) \qquad \text{by definition of } f$$

$$= \bar{\delta}_N((a, b_k), \bar{\lambda}_N((a_0, b_0) \cdots (a_{k-1}, b_{k-1}), p_0))$$

$$= \delta_N((a, b_k), p_k) \qquad \text{by definition of } p_k$$

Theorem IV

If M is embedded in N with q_0 embedded in p_0 relative to b_0, then M' as defined above with f_{b_0} is the reduced machine of M.

Proof:

Now for each state q in M either $q = q_0$ or $q = \lambda_M(a, q_0)$ for some a in ΣA. Hence, each f_q is either f_{q_0} or $f_{q_0} \circ L_a$ for some a in ΣA. Thus, the set of functions $\{f_q\}$ is completely determined by f_{q_0}, and so also is the reduced machine of M. Also, machine M' of Lemma 3 is determined from f_{b_0} in exactly the same manner as the reduced machine of M is determined from f_{q_0}. Thus, it suffices to show that f_{q_0} is the same function as f_{b_0} to prove M' is the reduced machine of M.

Let $a_0 \cdots a_n$ be any sequence in ΣA and $(a_0, b_0) \cdots (a_n, b_n)$ the associated allowable sequence for N.

Then $f_{q_0}(a_0 \cdots a_n) = \bar{\delta}_M(a_0 \cdots a_n, q_0)$ by definition of f_{q_0}

$= \bar{\delta}_N((a_0, b_0) \cdots (a_n, b_n), p_0)$ by Lemma 2

$= f_{b_0}(a_0 \cdots a_n)$ by definition of f_{b_0}

Proof of Theorem II: By Theorem IV, M' is the reduced machine of M, and by Theorem III, it is embedded in N.

Hence, while the machine M itself may not be constructed by the above procedure, a machine M' that in functional aspects is indistinguishable from M is constructed. As previously noted, even for a fixed initial state, different choices of b_0 may give rise to a different function, f_{b_0}, and hence a different machine. Similarly, a change of initial state may give rise to different functions.

A finite-state machine may be viewed as a set of functional transformations of a sequence of input symbols into a sequence of output symbols. These functions may be expressed in tabular form by relating to each input sequence that output symbol and next state which will result, given that the machine is in a particular initial state. Of course, it is only necessary to list those functions that give rise to different pairs of output symbol and next state when applied to sequences of length 1. Any state function is completely determined by the state and the associated output-symbol, since $f \circ L_1$ and $f \circ L_2$ with the same set of output-symbol and next-state pairs in both are the same function.

For example, if this procedure is applied to Machine N (Figure 3.24), then the set of embedded machines are obtained as follows. Starting in state A with a previous output-symbol 0, any input results in a transition to state B. In either case, the output is 1, so that the function

f (or, more exactly, f_0) is defined by the output-symbol next-state pairs $1B$, as shown in Table 3.8. If we follow this with an input of 0, the result is a transition to state A and an output of symbol 2. Similarly, an input of 1 results in the output-symbol next-state pair $0B$. These pairs, $2A$ and $0B$, correspond to $f \circ L_0$, as shown in the table. Note that these same pairs would arise for $f \circ L_1$, so that this function need not be shown. The remainder of the table is constructed in the same manner. Since there is only a finite number of output-symbol next-state pairs, there is only a finite number of different functions, $f \circ L_a$. Moreover, whenever the pairs that occur in the table for $f \circ L_a$ (where a is any input sequence of length k) have already occurred in the table for $f \circ L_b$ (where b is a sequence of length less than k), then all possible functions have been obtained.

Table 3.8

Starting in State A with a Previous Output of 0

Input	f	$f \circ L_0$	$f \circ L_{00}$	$f \circ L_{01}$	$f \circ L_{001}$	$f \circ L_{010}$
0	$1B$	$2A$	$1B$	$0C$	$3A$	$2B$
1	$1B$	$0B$	$1A$	$2B$	$0A$	$3C$

Table 3.9 is generated from the same machine but with the assumption of a different input symbol and initial state. Note that there is no need to generate the remainder of this table since the output-symbol next-state pairs for $f \circ L_0$ and $f \circ L_1$ are identical with those found in Table 3.8 so that the remainder of both tables must be the same. In other words, the machine expressed by Table 3.9 is the same as the machine expressed by Table 3.8, except for an extra initial state.

Table 3.9

Starting in State B with a Previous Output of 3

Input	f_1	$f_1 \circ L_0$	$f_1 \circ L_1$
0	$2B$	$2A$	$2A$
1	$2B$	$0B$	$0B$

Tables 3.10 and 3.11 are derived in a similar manner. In the latter case the machine has an extra initial state which, under 0, goes to the state $f \circ L_{00}$ of the machine expressed in Table 3.8 but, under 1, goes into itself.

Table 3.10
Starting in State C with a Previous Output of 1

Input	f_2
0	$2B$
1	$2B$

Table 3.11
Starting in State C with a Previous Output of 2

Input	f_3	$f_3 \circ L_0$	$f_3 \circ L_1$
0	$2A$	$1B$	$2A$
1	$2C$	$1A$	$2C$

A state diagram of the embedded machine $M(0A)$, as given in Table 3.8, can be generated by writing a state for each different left translate composition with f_0. (See Fig. 3.25.) The embedded machines obtained with the starting pairs $1A$, $2A$, $3A$, $0B$, $1B$, $2B$, $0C$ and $3C$ are the same as that obtained by starting with $0A$ since each of these pairs is contained in Table 3.8 and the machine is strongly connected. The starting pairs $3B$ and $1C$ give rise to a machine that is the same as the machine $M(0A)$ except for an extra state, which is also the initial state. The pair $2C$ gives rise to the embedded machine shown in Table 16 which is the same as $M(0A)$ except for an extra state which is also the start state. With $0A$ as the initial pair, the next state is 3, whereas under the input of 1 the next state is the initial state.

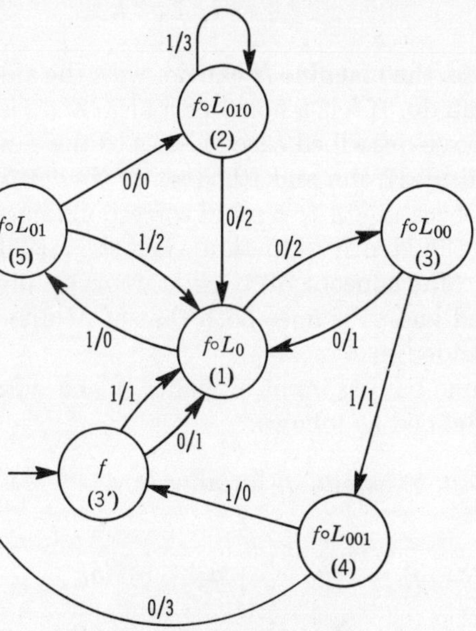

Fig. 3.25. Machine $(0A)$.

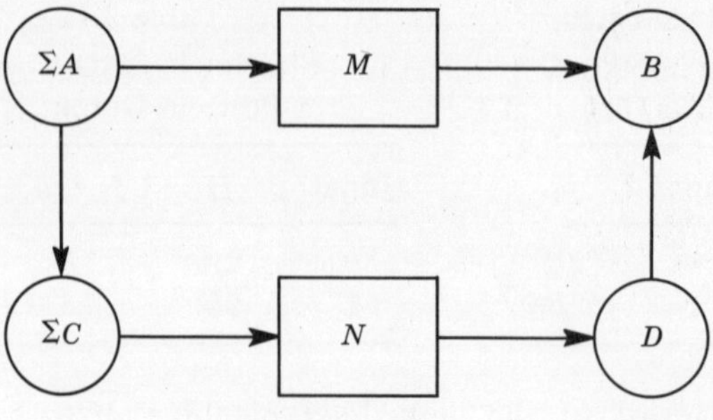

Fig. 3.26

Note that it is still possible that functions with the same output symbol but different next states may still be identical. Under such correspondence, these functions have to be compared on longer and longer sequences until either different output symbols result at some stage, or the output-symbol next-state pairs are found to be the same. In the first case, the functions are different; in the latter case, they are the same.

For machines $M = (A, B, Q, \lambda_M, \delta_M, q_0)$ and $N = (C, D, P, \lambda_N, \delta_N, p_0)$, suppose, as indicated in Figure 3.26, that there exist functions h and t such that

$$\delta_M = t \circ \delta_N \circ h$$

Then, in a sense, the machine N can do, with the aid of these functions, whatever M can do. If h is a homomorphism and t is any function from D to B, the above-described case reduces to the concept of M dividing N (as proposed by Krohn and Rhodes). If $C = A \times B$, h is a function from the sequences in ΣA to a set of corresponding allowable sequences in ΣC and t is the identity function, then the machine M is embedded in machine N. The concept of M being functionally contained in N is now introduced and subsumes both that of M dividing N and that of M being embedded in N.

For a machine having input alphabet A and output function δ, the function $\hat{\delta}$ is defined as follows:

$$\hat{\delta}(a, q) = \delta(a, q) \text{ for all } a \text{ in } A,$$

and

$$\hat{\delta}(a_0 \cdots a_n, q) = \hat{\delta}(a_0 \cdots a_{n-1}, q)\bar{\delta}(a_0 \cdots a_n, q),\ a_i \text{ in } A$$

that is, $\hat{\delta}(a_0 \cdots a_n, q)$ is the output sequence when the input sequence is $a_0 \cdots a_n$ and the machine is in state q.

Consider the following three classes of functions:

1. The set of all homomorphisms on input semigroups to input semigroups, where an input semigroup is a free semigroup on a finite alphabet.

2. The set of all left translates on these input semigroups.

3. The set of functions of the form,

$$
\begin{aligned}
{}_1\hat{\delta}(a_0 \cdots a_n, q) &= \hat{\delta}(a_0 \cdots a_{n-1}, q) && \text{if } n \geq 1 \\
&= \phi(\text{the null sequence}) && \text{if } n = 0
\end{aligned}
$$

Let H be the set of functions generated from these three sets by means of the operations of composition and direct product; that is, H is the smallest set of functions that contains those functions in the above sets 1, 2, and 3, and which is closed under the operations of composition and direct product. For $M = (A, B, Q, \lambda_M, \delta_M, q_0)$ and $N = (C, D, P, \lambda_N, \delta_N, p_0)$, M is defined as being functionally contained in N (denoted $M < N$) if there is an h in H and a function t mapping D into B such that $\delta_M = t \circ \delta_N \circ h$ where δ_M and δ_N have as initial states q_0 and p_0, respectively.

Now H contains the set of input semigroup homomorphisms; hence, if M divides N, then M is functionally contained in N. To show that M embedded in N implies functional containment, consider the cartesian product of the identity function on ΣA and the function $L_{b_0} \circ {}_1\hat{\delta}_M$. This is readily seen to be the function of ΣA into $\Sigma(A \times B)$ required for machine embedding. Since the identity function and the function L_{b_0} and ${}_1\hat{\delta}_M$ are in H, the embedding function being derived from these by composition and cartesian product is likewise in H.

Since the identity function on a semigroup is a homomorphism and H is closed under composition of functions, the relation "functionally contained in" is reflexive and transitive. Let M be functionally equivalent to N (denoted $M \sim N$) if both $M < N$ and $N < M$ occur. Then $\sim$ is an equivalence relation and partitions the set of finite-state machines into classes of functionally equivalent machines. The relation $<$ is induced on the equivalence classes in that for any two equivalence classes M and N, there is an M in M and an N in N such that $M < N$, if and only if every machine in M is functionally contained in every machine in N.

Theorem IV

If $M < N$ and M_1 is derived from M by changing the initial state from q_0 to q_1, then $M_1 < N$.

The required function from H is given by $h \circ L_a$ where $q_1 = \tilde{\lambda}_M(a, q_0)$ and h is the member of H under which $M < N$. Since $M < M$, the following corollary results: If M_1 is derived from M by changing the initial state, then $M_1 < M$.

Further, since in a strongly connected machine M, each state can be reached from every other state, each machine derived by a change of initial state is equivalent to M, and hence to every such derived machine.

Theorem V

If h is in H, mapping ΣA into ΣC, then for a in ΣA such that $a = a_1 a_2$, $h(a) = h(a_1)c$; that is, h is a quasi-homomorphism.

To prove this, it suffices to note (1) that functions in the basic classes defining H are quasi-homomorphisms, (2) that the composition of two quasi-homomorphisms is a quasi-homomorphism and (3) that so also is the direct product.

If K is the set of quasi-homomorphisms on input semigroups to input semigroups, then, by Theorem V, H is contained in K. Why not replace H by K in the definition of one machine being functionally contained in another? If this is done, then all of the already obtained results can be derived. Unfortunately, some unwanted results are also derivable. For example, the function $\hat{\delta}$ is in K. Consider then any machine $M = (A, B, Q, \lambda_M, \delta_M, q_0)$ and the corresponding one-state throughput machine $N = (B, B, \{p\}, \lambda, \delta, p)$ where δ is the identity function. Then with the function $\hat{\delta}_M$ as the required quasi-homomorphism, M is functionally contained in N. Hence, the set K is too large to yield a reasonable theory since using it implies that every machine is functionally contained in a one-state machine.

In retrospect, the embedding technique permits the translation of an evolved machine into the set of possible reduced machines that are embedded within it, so that these may be viewed as reasonable hypotheses concerning the logical relationship between the sensed variables.

CHAPTER FOUR

Data Reduction and Analysis

DIAGNOSIS, DETECTION, AND DISCRIMINATION

As indicated in Chapter 2, evolutionary programming is suitable for diagnosis of the environment (interdiction). Consider the situation wherein a number of physiological measurements are made on a patient each day. On the basis of these measurements and past diagnoses, the doctor is required to identify the current status of the patient's health. To illustrate, let the first parameter have two possible symbols, while the second and third are each specified within four symbols. On each day the status of the patient's health can be described by one of 5 symbols. Let 0 represent a serious disorder that may directly cause death and let 1 represent a serious disorder but one not usually fatal. The three remaining symbols portray situations of lesser significance. Specifically, let 2 represent any disorder which is not serious but cannot be immediately cured by treatment. The number 3 represents any disorder which is not serious but can be cured immediately by treatment, and 4 represents the state of perfect health. A fictitious data base was generated, covering a time period of about 3 years. Although these data were not taken from medical records, they provide an interesting exercise, one that demonstrates the capability of evolutionary programming, even within a severely restricted alphabet.

Because the available program is restricted to an 8-symbol alphabet, it was only possible to perform experiments using two of the three sensed parameters in combination. The original intent of the available program was to predict each next symbol in the input sequence. To make this program perform the desired diagnosis, it was necessary to modify the internal comparison so as to replace each next input symbol with the event symbol (the status of current health); this latter variable being shifted ahead one unit of time.

The first experiment was conducted in order to demonstrate that interdiction was being properly performed. For this experiment the data base contained only those days during which some activity was observed (this in terms of the occurrence of a symbol other than 0 on any of the observed parameters). A penalty of 0.01 per state was used in order to ensure that the evolved finite-state machines remain of reasonable size. The input variable was generated by combining parameters 1 and 4, in that order, with the result being coded, as shown in Table 4.1. Thus, the output variable was made to appear as a portion of the input variable. Perfect interdiction is recognized by this correspondence.

Table 4.1

Input Data Coding

Uncoded	Coded
0 0	0
0 1	1
0 2	2
0 3	3
1 0	4
1 1	5
1 2	6
1 3	7

The first evolved machine was of five states, this being a different machine than the arbitrary original five-state machine. After the comparative evaluation of 21 offspring, the best evolved machine was a two-state machine. At the 178th decision, the best machine was still of two states and the logic of the coding was becoming apparent. Thereafter, the best evolved machine had only one state, this closely reflecting the code. After the first 225 decisions, all but one of the state transition output symbols within the evolved machine were correct, this one having never been exercised by the input data.

The second experiment was concerned with the identification of the concurrent event based upon the available history of parameters 1 and 2. The same coding and program conditions were used as in the previous experiment, but the error cost matrix was allowed to be an overly simple expression of the analyst's goal. The relative worth of each correct interdiction was placed on the main diagonal. Thus, no penalty is given for a correct interdiction of 0 and a penalty of one unit accompanies a correct interdiction of 1. The lesser importance of symbols 2, 3, and 4 is shown by their weightings of 3, 4 and 5, respectively. All off-diagonal terms were arbitrarily taken to be of weight 9 in order to equally discourage all possible errors. In this experiment, 1,141 different offspring were evaluated in the process of making the first 301 decisions, the last of these being based on a twenty-one state machine. On the average, 62.1 percent of these decisions were correct.

With this as encouragement, it was considered appropriate to refine the error cost matrix in order to more fully express the goal. All main-diagonal terms were given a weight of 0 while the off-diagonal terms were given weights that express the relative cost of making the associated error. More specifically, each element in position ij of the error cost matrix is a measure of the relative penalty for interdicting a j when the actual symbol is an i. For example, if i is of greater importance than j then the ij entry is made greater than the ji entry in order to show this preference. As a result, the sum of the elements in a column that corresponds to a symbol of lower importance is greater than the sum of the elements in a column that correseponds to a symbol of greater importance. In addition, the elements in the 0 column were given a slight increase in weight in order to reduce the number of times that this symbol will appear as the result of a decision; this is meant to overcome the prevalence of 2's in the sequence of events. The error cost matrix shown in Table 4.2 was taken to represent a more realistic statement of the goal to be achieved.

Table 4.2

Error Cost Goal Matrix for Experiment 3

		Interdicted							
		0	1	2	3	4	5	6	7
	0	0	4	6	6	7	9	9	9
	1	3	0	5	5	5	9	9	9
	2	1	1	0	2	3	9	9	9
Actual	3	2	2	2	0	2	9	9	9
	4	2	2	3	2	0	9	9	9
	5	9	9	9	9	9	0	9	9
	6	9	9	9	9	9	9	0	9
	7	9	9	9	9	9	9	9	0

The third experiment was identical with the second experiment except for the substitution of this new error cost matrix. Here, 1,171 different offspring were evaluated in the process of making the first 301 decisions, the last of these being based on a nine-state machine. On the average 61.1 percent of these decisions were correct. This slight decrease in percent correct score can be expected, in view of the greater specificity of the new goal statement. A more meaningful view of the result of this experiment can be gained by examining the relative number of each correct and each incorrect type of decision in, say, the last 100 decisions. These data are shown as percentages in Table 4.3. That is to say, 90 percent of the 0's experienced were properly interdicted, while 10 percent of these 0's were incorrectly interdicted as being 2. Clearly the most serious problem is associated with interdiction of symbol 1. This, together with the excessive number of 2's which were interdicted (45 in the 100 decisions), justified a revision of the error matrix (see Table 4.4).

Table 4.3

Percentage Results of Experiment 3

		Interdicted			
		0	1	2	3
Actual	0	90.0	0	10.0	0
	1	59.1	18.2	13.6	9.1
	2	11.1	31.1	57.8	0
	3	33.3	0	66.6	0

Table 4.4

Error Cost Goal Matrix for Experiment 4

		Interdicted							
		0	1	2	3	4	5	6	7
Actual	0	0	4	6	6	7	9	9	9
	1	4	0	5	5	5	9	9	9
	2	1	2	0	2	3	9	9	9
	3	2	2	2	0	2	9	9	9
	4	2	2	3	2	0	9	9	9
	5	9	9	9	9	9	0	9	9
	6	9	9	9	9	9	9	0	9
	7	9	9	9	9	9	9	9	0

In this experiment 1,021 different offspring machines were evaluated in the process of making the first 301 decisions, the last of these being a fifteen-state machine. On the average, 66.5 percent of these decisions were correct, but the results of this fourth experiment are more meaningfully expressed in Table 4.5. Note the improvement in the interdiction of 1, 2, and 3, this at a small cost to the correct interdiction of 0. The same number of 2's appeared in the last 100 decisions.

Table 4.5
Percentage Results of Experiment 4

		Interdicted			
		0	1	2	3
Actual	0	83.4	0	16.6	0
	1	45.5	18.2	31.8	0
	2	9.0	8.9	78.8	2.2
	3	0	0	66.6	33.3

Table 4.6
Percentage Results of Experiment 5

		Interdicted			
		0	1	2	3
Actual	0	83.4	0	16.6	0
	1	63.6	4.5	31.8	0
	2	11.1	24.4	62.3	2.2
	3	66.6	0	33.3	0

Even though further improvement on the interdiction of the event variable based on parameters 1 and 2 would appear possible, it was considered of greater interest to attempt to interdict this same variable based on parameters 1 and 3. The fifth experiment was carried out using the last described error cost matrix as the statement of goal. In this experiment, 1,041 different offspring were evaluated in the process of making the first 301 decisions, the last of these being based on a fourteen-state machine. On the average, 65.8 percent of these decisions were correct. Table 4.6 reveals that these results are comparable for symbols 2 and 0 but somewhat poorer for symbols 1 and 3. In spite of the fact that some better error cost matrix might be found for this interdiction problem, it is reasonable to conclude that parameter 2 is of greater worth than parameter 3 in terms of the goal defined by the error cost matrix of Table 4.4.

It is of interest to attempt to predict the next event at each point in time. Therefore, an experiment was conducted in which each next event was predicted on the basis of only the preceding sequence of events. A total of 1,811 offspring were evaluated in making the first 301 predictions, the last being made by an eighteen-state machine. Of these predictions, 48 percent were correct (see Table 4.7, which shows that 70 percent of the most significant events that occurred were, indeed, properly predicted).

Table 4.7

Percentage Results of Experiment 6

		Predicted			
		0	1	2	3
Actual	0	70.0	6.7	23.3	0
	1	72.7	4.6	22.7	0
	2	40.0	20.0	40.0	0
	3	0	0	100.0	0

For the sake of comparison, programs were written which would predict these same data in terms of that symbol which had occurred most frequently, that symbol which was associated with the largest first-order conditional probability, and that symbol which was associated with the largest second-order conditional probability, respectively. Of these, the zeroth-order predictor proved to be slightly superior to the other two, reaching a score of 50 percent; however, it is worthwhile to note that this score corresponds with the percentage of 2's in the event sequence. None of the 0's, 1's, and 3's which occurred were correctly predicted. The first-order predictor demonstrated a final score of 49.3 percent, having properly predicted two of the thirty 0's which appeared in the data. Although the second-order predictor attained a score of 47.3 percent and showed some tendency to predict symbols other than 2, it also properly predicted only two of the thirty 1's which appeared in the data. Clearly, these conventional techniques fall far short of the ability demonstrated by the evolutionary program.

The experiments just described were conducted on the abridged data base (that includes only days when the patient is noted to be disturbed). As a result, such experiments could not be expected to reveal purely temporal patterns which might exist within the data. It is therefore of interest to predict each event, based upon an unabridged combination of parameters 1 and 2. In Experiment 7, some 1,311 different offspring were evaluated in making the first 400 predictions, the last of these machines being of four states. Some 67.5 percent of these predictions were correct. Table 4.8 provides a more meaningful breakdown of these results. The greatest percentage of 2's, 0's, and 4's were correctly predicted. Only a single 1 appeared within the last 100 events. It is apparent that the error cost matrix properly reflects not only the relative worth of each event and each of the possible incorrect predictions but also a suitable bias that precludes the constant prediction of symbol 4 (perfect health), by far the most frequent symbol.

Table 4.8

Percentage Results of Experiment 7

		Predicted				
		0	1	2	3	4
Actual	0	57.1	0	14.3	0	28.6
	1	0	0	100.0	0	0
	2	0	11.1	61.1	0	27.8
	3	0	0	0	0	100.0
	4	0	0	10.5	0	80.5

A comparable experiment was then performed in which each next event was predicted only on the basis of the previous daily history of events. The first 396 predictions were based on an evaluation of 2,471 different offspring machines, the last of these being of 21 states. 37.6 percent of these predictions proved to be correct. A more detailed analysis of these results is shown in Table 4.9. These poor results demonstrate the value of using the sensed parameters to enhance the predictability. Presumably the use of the history of events combined with data on the three sensed parameters should provide even better prediction than that achieved through the use of either of these classes of data alone.

These experiments are considered to demonstrate the feasibility of using evolutionary programming as a means for diagnosis and prediction of real-world data.

Two of the fundamental problems of information technology, already considered in Chapter 3, are worthy of restatement. Situations often arise in which a sequence of measurements are examined in order to determine whether or not any pattern or regularity exists within the data. An analogous situation is found in the first step of cryptanalysis: "Is a message contained in this sequence of symbols?" The problem of *detection* can be translated into the problem of finding sudden changes in the predictability of the data. A consistent inability to predict each next symbol leads to the suspicion that it is purely random. A steady and significant prediction score reveals the existence of an unchanging signal. In contrast, variability of the prediction score reveals that the data may contain a message.

Table 4.9

Percentage Results of Experiment 8

		Predicted				
		0	1	2	3	4
Actual	0	42.9	0	14.3	14.3	28.6
	1	0	0	0	0	100.0
	2	29.7	13.5	18.9	2.7	35.1
	3	0	0	8.3	16.7	75.0
	4	27.2	6.9	11.6	0	44.2

Demonstration of an ability to predict the data immediately leads to the question: "What is the nature of the signal?" The answer to this question may be found in terms of the evolved predictor-machines. Each machine may be viewed as a description of either periodic or stochastic properties of the environment. As previously indicated, every finite-state predictor-machine has a characteristic cycle which can be found by driving the machine by the sequence of its own output symbols. At some point, the sequence of output symbols will become an endless repetition of the characteristic cycle. If the environment contains periodic properties these should be found in the characteristic cycle.

At the same time each finite-state machine may be viewed as that deterministic procedure which best typifies the statistical properties of the data. If the environment is considered to be of first order, each state transition should correspond with the probabilistic transitions which occur with highest conditional dependency. For example, as discussed in the previous chapter, Experiment 11 was an attempt to predict the 4-symbol environment generated by an arbitrary transition matrix. The transitions specified by the last evolved predictor-machine correspond with those of the greatest first-order conditional probability for each input symbol. By the same token, the zeroth-order properties of the environment should be evident in terms of the relative frequency of the output symbols of the envolved predictor-machines. It is more difficult to estimate the second- and higher-order conditional probabilities of the environment directly from the evolved finite-state machine. In a sense, this information is stored in the machine because it has effected the sequence of selected machines; however, direct read-out of this information is not always possible. At one extreme, a repetitive environment may be viewed as the limiting case of a stochastic process in which the order equals the length of the period. Under such conditions the evolved machine will be cyclic and the "conditional" dependency is clearly evident. At the other extreme, where the conditional probabilities are nearly equal, it becomes difficult if not impossible to estimate these from the configuration of the evolved machine. Of course, if such information appears to have particular value, the evolutionary program could be written to consider extended finite-state machines, machines that designate each transition in terms of the last few symbols, the number of these defining the highest-order conditional properties to be explicitly expressed.

All too often the search for artificial intelligence is reduced simply to attempts to build automatic pattern recognition devices. Seen in context, this is but one of the capabilities that identify the existence of intellect. First, there is the problem of detection: "Is there a signal in the noise?" Then there is the problem of discrimination: "If there is a signal, what is it?" With this accomplished, it now becomes reasonable to inquire: "Is this signal one of those already known?" Quite literally, the problem is one of re-cognition. In the usual situation, the problem is posed in a larger sense: "To which of the known signals is this signal most similar?" This is the problem of pattern classification.

PATTERN RECOGNITION AND CLASSIFICATION

Begin with this question: "Can an evolutionary program recognize and classify patterns in a manner similar to that of an average human operator?" To make the question specific, sixteen different amplitude-time patterns were generated, these naturally falling into four classes in terms of their similarity. Each pattern was generated in terms of 35 possible amplitudes at each of 125 equally separated points in time by recording the rectified envelope of a waveform that resulted from

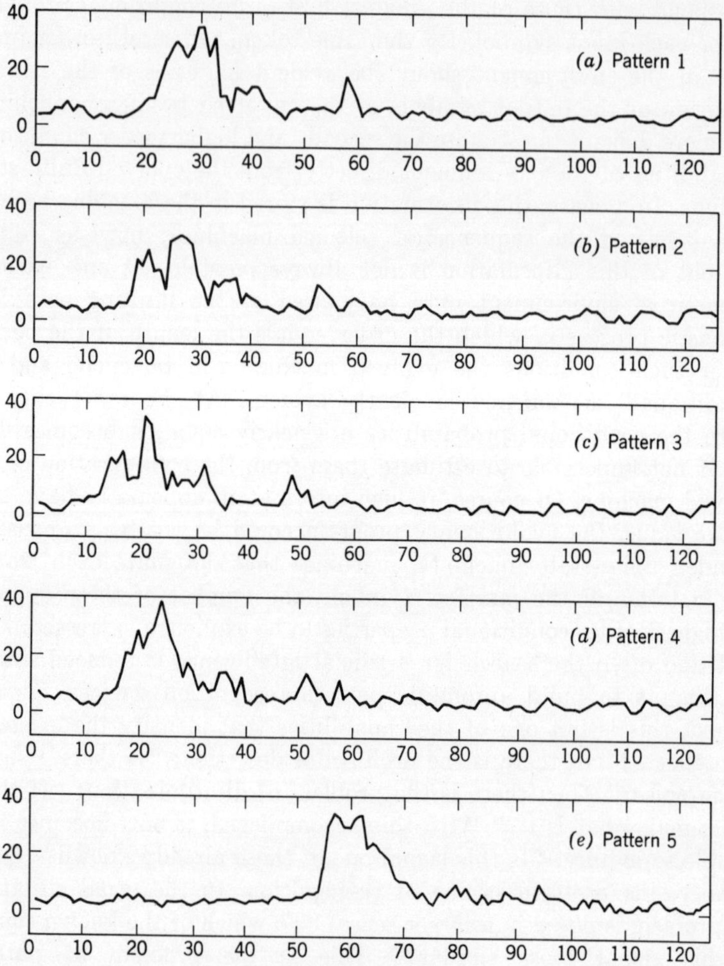

Fig. 4.1

adding Gaussian noise to four different primary events. (See Figures 4.1*a* to 4.1*p*.)

In order to make these data suitable for the available computer program, the amplitude data were encoded into an 8-symbol alphabet, this in a manner that would tend to equalize the marginal probabilities. For the sake of efficiency, each pattern was begun with the first excursion from the symbol 0 and terminated after 45 symbols had appeared. The encoded data are shown in Figures 4.2*a* through 4.2*p*.

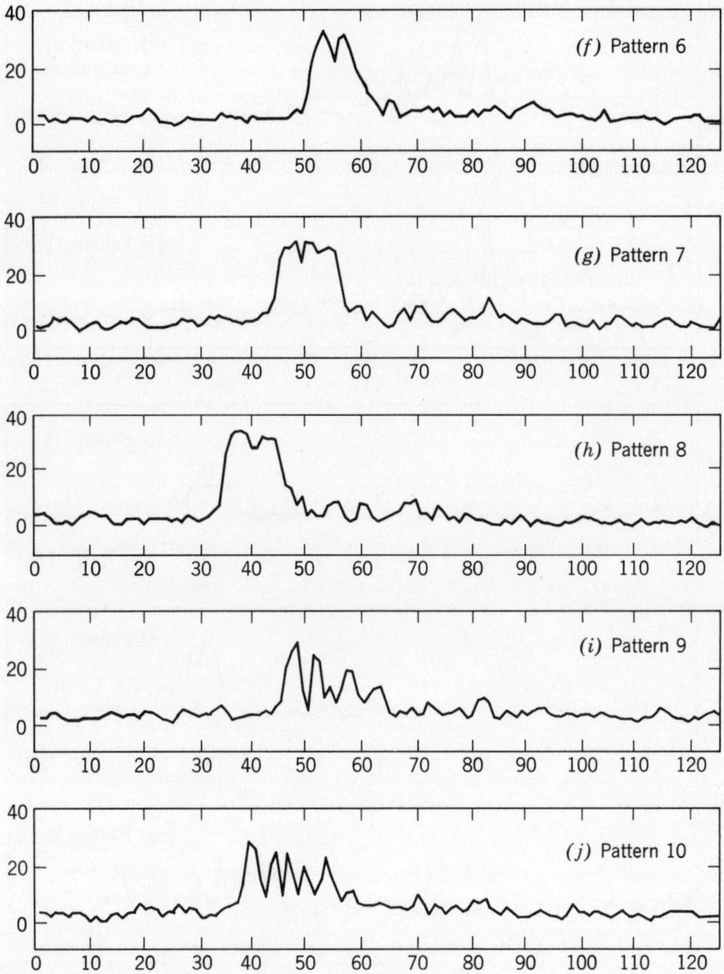

Fig. 4.1 (*continued*)

Note that some degree of subjective similarity remains among the members of each successive quadruplet.

The previously described 8-symbol evolutionary program was used to predict each next symbol in an unending repetition of each of these patterns, there being no penalty for complexity, ten generations prior to each prediction, and a "magnitude of the difference" error cost matrix specification of the goal. Table 4.10 indicates the average error in predicting the past for each of the predictor-machines after the

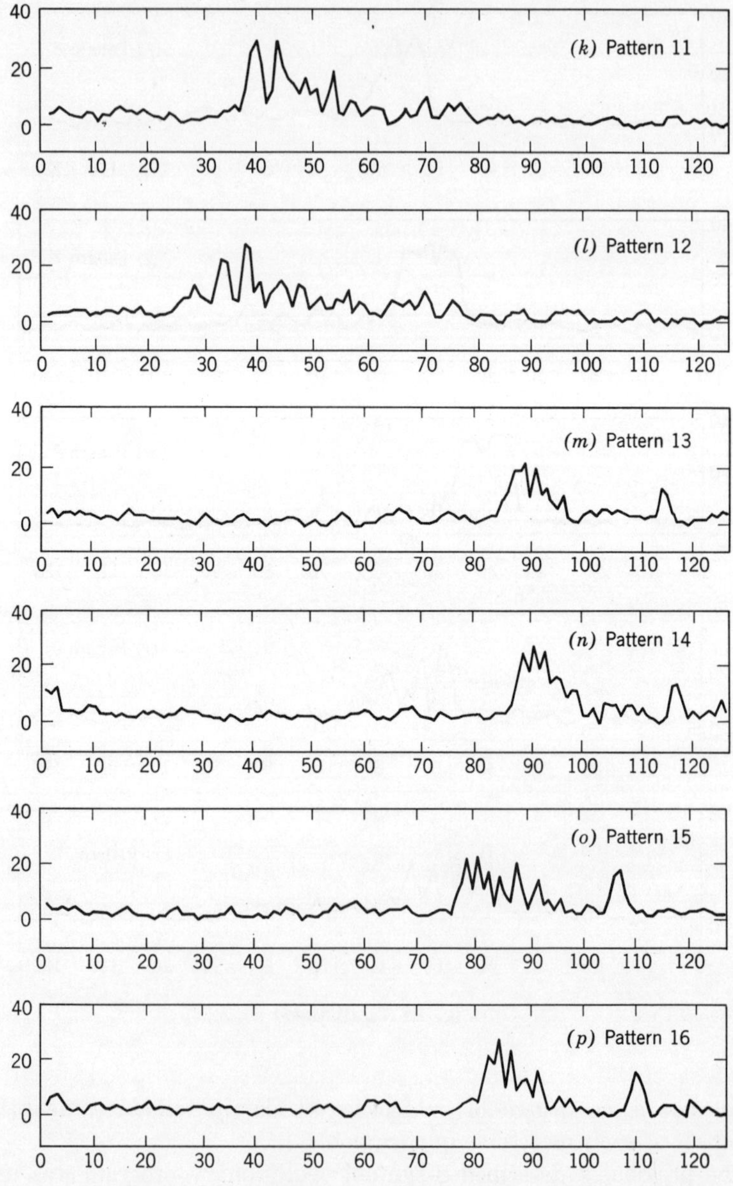

Fig. 4.1 ***(continued)***

first 50, 100, 200, and 400 predictions. The greatest portion of the learning is seen to take place early in the evolution.

Each of the evolved predictor-machines is a particular characterization of the pattern against which it was evolved. Similarity of the patterns might be deduced by noting similiarties in these machines; however, their complexity would make this difficult even if a suitable set of parameters were available. A more natural way to accomplish this same comparison is to allow each evolved predictor-machine to

Table 4.10

	Recall Length			
Pattern	50	100	200	400
1	0.330	0.172	0.158	0.165
2	0.307	0.306	0.211	0.145
3	0.228	0.195	0.091	0.056
4	0.248	0.209	0.111	0.114
5	0.106	0.091	0.057	0.045
6	0.043	0.051	0.051	0.047
7	0.082	0.041	0.048	0.050
8	0.268	0.267	0.288	0.276
9	0.185	0.163	0.172	0.149
10	0.207	0.141	0.088	0.070
11	0.184	0.082	0.078	0.052
12	0.312	0.107	0.090	0.072
13	0.122	0.102	0.091	0.081
14	0.103	0.126	0.110	0.109
15	0.165	0.176	0.152	0.136
16	0.063	0.068	0.067	0.065

predict the remaining patterns. Hopefully, the similarity among certain patterns will be evident in terms of the nearness of their prediction scores. Each predictor should score highest on that pattern against which it was generated, nearly as high on "similar" patterns and very low on all other patterns.

In the next set of experiments one of the predictor-machines evolved against patterns 1, 2, and 3 was used to predict each of the sixteen patterns using a squared error cost matrix. Table 4.11 indicates the results with the average error of prediction tabulated over 45 and 90 symbols. As expected, there is little difference between these sets of data. Each predictor-machine recognizes its own pattern with ease but

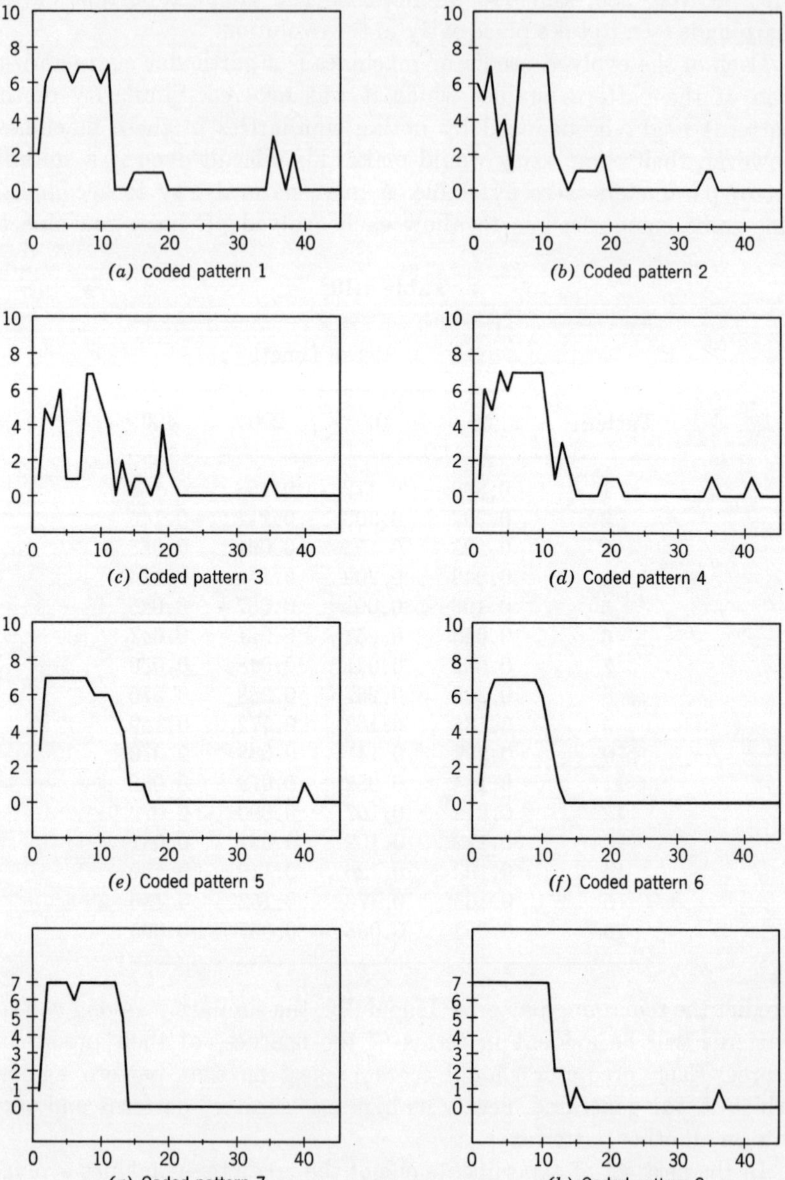

(*a*) Coded pattern 1
(*b*) Coded pattern 2
(*c*) Coded pattern 3
(*d*) Coded pattern 4
(*e*) Coded pattern 5
(*f*) Coded pattern 6
(*g*) Coded pattern 7
(*h*) Coded pattern 8

Fig. 4.2

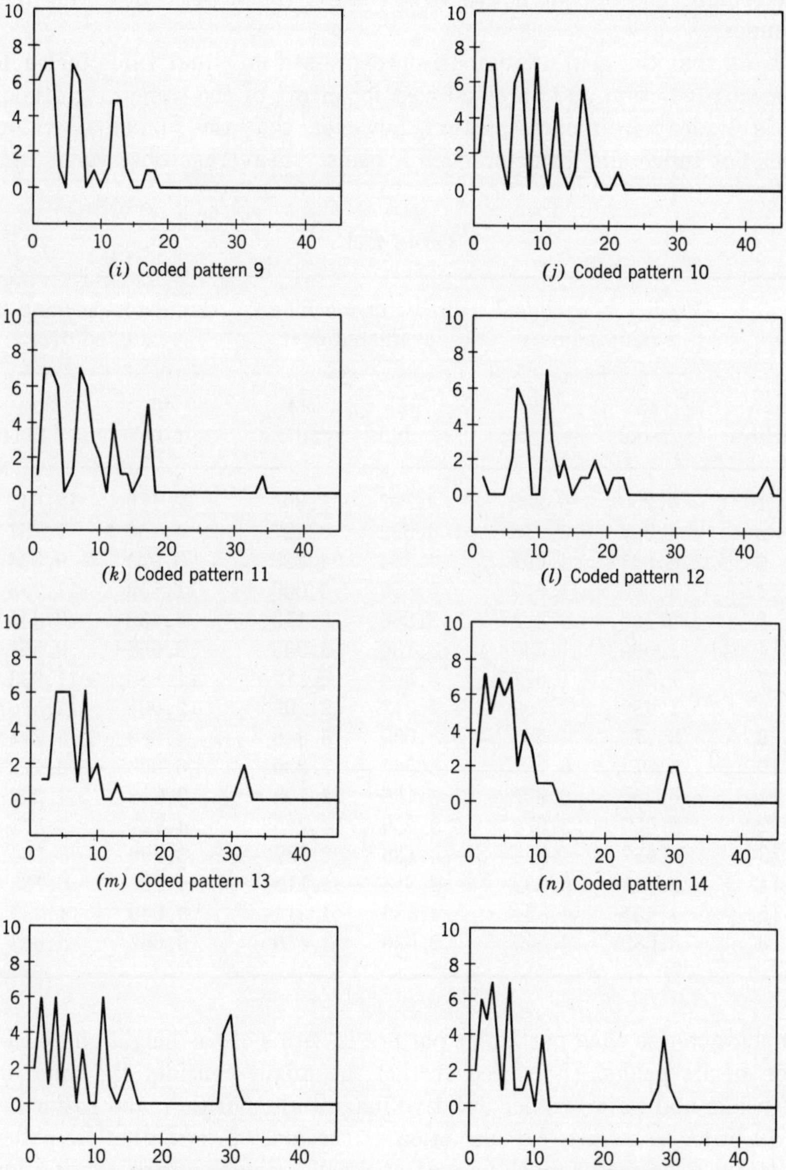

Fig. 4.2 *(continued)*

the remaining scores do not serve to classify the patterns in the desired manner.

Note that the goal of comparison expressed an equal value for each correct prediction and an equal cost for errors of the same magnitude. It is evident from introspection, however, that the human operator does not judge similarity on such a basis. An average observer is apt

Table 4.11

	Predictor-machine 1 evaluated over		Predictor-machine 2 evaluated over		Predictor-machine 3 evaluated over	
Pattern	45 symbols	90 symbols	45 symbols	90 symbols	45 symbols	90 symbols
1	0.250	0.292	4.727	5.865	12.136	12.045
2	4.727	5.315	0.023	0.427	5.159	7.787
3	4.841	4.697	4.864	4.820	0.023	0.034
4	4.500	3.629	5.659	6.000	11.159	11.258
5	2.796	3.348	5.886	5.112	5.432	8.843
6	1.091	1.539	3.136	3.933	9.068	9.551
7	1.296	1.697	3.364	3.112	11.886	11.933
8	1.432	2.753	2.318	2.292	12.091	12.506
9	3.773	4.685	5.000	5.348	4.682	5.315
10	6.273	6.787	8.546	7.225	6.068	6.472
11	5.250	5.607	5.136	4.798	3.818	4.292
12	3.000	2.978	3.250	3.225	4.727	4.573
13	4.432	4.258	3.136	3.337	3.796	3.764
14	5.341	5.382	3.818	4.112	6.727	6.753
15	4.455	4.854	4.386	4.214	5.046	4.989
16	3.682	4.090	2.636	3.416	5.682	5.663

to characterize each particular pattern in terms of the height and position of its peaks. If pressed further, he might consider the number of peaks and valleys, their relative magnitude, position, and so forth.

In order to ensure the extraction of the most information from the pattern, a predictor-machine was evolved against pattern 1 using an all-or-none goal (0's down the main diagonal, all off-diagonal terms being 9), a penalty of 0.01 per state; and a long initial recall (225 symbols). The resulting predictor-machine is shown in Table 4.12. This machine was then used for the predictive comparative evaluation of all encoded patterns using goals that placed increased value on

Table 4.12

Pattern 1*A*	
Error Score 1.0056	
Initial State	2
Present State	1 1 1 1 1 1 1 1 2 2 2 2 2 2 2 2 3 3 3 3 3 3 3 3 4 4 4 4 4 4 4 4
Input Symbol	7 6 5 4 3 2 1 0 7 6 5 4 3 2 1 0 7 6 5 4 3 2 1 0 7 6 5 4 3 2 1 0
Next State	3 1 3 2 1 3 4 2 2 2 1 1 2 3 4 2 1 4 1 2 1 2 1 3 3 4 1 2 3 3 1 3
Output Symbol	6 2 2 2 2 2 1 0 2 2 2 2 1 6 0 0 7 7 2 2 2 0 1 0 7 2 2 2 2 2 0 2

the correct prediction of "peak" (Table 4.13) and on "peaks and valleys" (Table 4.14). These results are shown in Tables 4.15 and 4.16, respectively. It would appear that some classification is taking place. Note the similarity of scores on patterns 5, 6, and 7, but this is hardly enough. There is a considerable difference between describing

Table 4.13

		Predicted							
		0	1	2	3	4	5	6	7
Actual	0	7	8	9	10	11	12	13	14
	1	8	6	7	8	9	10	11	12
	2	9	7	5	6	7	8	9	10
	3	10	8	6	4	5	6	7	8
	4	11	9	7	5	3	4	5	6
	5	12	10	8	6	4	2	3	4
	6	13	11	9	7	5	3	1	2
	7	14	12	10	8	6	4	2	0

Table 4.14

		Predicted							
		0	1	2	3	4	5	6	7
Actual	0	0	1	2	3	4	5	6	7
	1	2	1	2	3	4	5	6	7
	2	4	3	2	3	4	5	6	7
	3	6	5	4	3	4	5	6	7
	4	7	6	5	4	3	4	5	6
	5	7	6	5	4	3	2	3	4
	6	7	6	5	4	3	2	1	2
	7	7	6	5	4	3	2	1	0

Table 4.15

Pattern	Evaluated over 45 symbols	Evaluated over 90 symbols
1	5.6136	5.6517
2	6.8182	6.8427
3	7.4773	7.4831
4	6.5455	6.6854
5	6.2273	6.2584
6	6.2273	6.2472
7	6.0000	6.0225
8	7.7955	7.8652
9	7.3864	7.2921
10	7.5909	7.6067
11	7.2955	7.3034
12	7.6364	7.4494
13	7.6364	7.6404
14	7.3409	7.2921
15	7.6818	7.8090
16	7.4773	7.6742

Table 4.16

Pattern	Evaluated over 45 symbols	Evaluated over 90 symbols
1	0.5909	0.6292
2	1.2727	1.3258
3	1.5682	1.5730
4	1.0909	1.1685
5	0.8636	0.8989
6	0.5682	0.5843
7	0.5455	0.5618
8	1.4773	1.5393
9	1.0682	1.0899
10	1.5000	1.5281
11	1.2727	1.2809
12	1.2500	1.1685
13	1.1136	1.1236
14	1.2955	1.3146
15	1.4773	1.5393
16	1.2727	1.3820

a given pattern *as a pattern* and describing a pattern *as compared to a set of other patterns*. The given problem falls in this latter class. It would therefore be necessary to use a payoff matrix that includes such features as degrees of "choppiness" and "relative peak height." This precludes the use of a point by point evaluation technique that by its very nature excludes such "gestalt."

To summarize, a number of experiments were conducted. It was immediately evident that pattern recognition was readily achieved in terms of the evolved predictor-machines; however, pattern classification was not accomplished in a manner similar to that of an average human observer. It is also evident that the cost matrix which expresses the goal of comparison must embody the characteristics that are the basis for human judgment of similarity. Of course, if these parameters were already known there would be no need for evolving predictor-machines. Such knowledge could be more fruitfully used in terms of a heuristic program. But in many situations it is of far greater importance to detect similarities that lie beyond human ability to recognize (as, for example, in sonar target classification and camouflaged target recognition). Here, the problem is to identify some underlying similarity *not* evident to the observer. It is considered worthwhile to attempt to use evolutionary programming as a technique for the solution of such problems. Presumably, a set of properly classified samples can furnish a basis for determining the goal that expresses the desired classification.

CHAPTER FIVE

Control System Design

CONTROL OF AN UNKNOWN PLANT

Chapter 4 was concerned with such problems of data reduction as *detection* (is there a signal?), *discrimination* (if so, what is the signal?), *recognition* (has this signal been seen before?), and *classification* (if not, which of a set of signals is it most like?). Only at this point is it appropriate to address a more significant problem: *the problem of control.* (A situation has been characterized, now what should be done about it?) In essence, the previous problems were of interest only in that they might precede steps toward a solution of the problem of control.

Situations often arise in which it is desired to make a given system (here referred to as the plant) perform in a particular manner. If this desired behavior is not expected to result from the natural interaction of the plant with its environment, the problem is to manipulate the available input parameters of the plant in a manner that will cause this performance, in spite of the essential dynamics of the plant and any internal or external disturbances that may occur. The problem is made more severe if knowledge of the nature of the plant and the disturbance is incomplete. Moreover, the situation becomes extremely difficult if the desired performance can only be measured when it is nearly complete, or if the evaluative criteria have been only partially specified.

This very general view of the control system design problem is offered only to emphasize the limitations of the current state of the art. Conventional techniques can be used if it is desired to make a linear plant satisfy certain constraints on its steady-state or transient response. Methods are currently available for the design of control systems wherein the plant possesses only certain kinds of nonlinearity.

To date, little attention has been devoted to the design of control systems for distributed plants (plants that require an infinite number of descriptors to fully define their state). Some of the effects of stochastic disturbance have been examined in considerable detail, but the effective control system design procedures reduce such problems to their deterministic "equivalent" by considering the average disturbance.

Analytical design procedures become inadequate in the face of incomplete knowledge of the plant to be controlled. Iterative techniques may be called upon, but their success depends upon the availability of suitable evaluation criteria that can be referenced at each iteration. The problem becomes even more formidable when the performance of the plant can only be evaluated at particular points in time. In the limit the problem becomes unsolvable as the time remaining to achieve control runs out.

For the sake of this discussion, attention will be restricted to plants that operate only at discrete points in time and have input parameters and performance that can be measured within some finite set of symbols. The parameters are taken collectively to comprise the input variable. The performance is described in terms of a sequence of symbols that constitute the controlled variable. The plant is presumed to be in one of some finite number of discriminably different internal states during each transduction. The worth of each symbol of the controlled variable is measured in terms of a completely specified payoff matrix that relates actual performance to desired performance at each point in time. The elements of this matrix may be time dependent or may even be functions of the worth of previous performance. The desired overall performance may be that which will maximize the worth over some sequence of symbols received from the environment.

Suppose the intent is to have a given, but unknown, plant optimally track an arbitrary reference signal in the sense of satisfying any well-defined goal. An evolutionary control system to accomplish this end operates in the following manner: The sequence of symbols which has most recently driven the given plant, M, is used to drive an arbitrary initial finite-state machine M_1'. Each output symbol of M_1' is compared with the corresponding symbol that had emerged from M. The penalty for any discrepancy is then found by referencing the error cost matrix that is an explicit statement of the purpose of the control system. The cumulative penalty over the sequence of symbols that comprises the recall provides a measure of the inadequacy of M_1' as a representation of M in the light of this goal and the experienced environment.

An offspring of M_1' is now produced through mutation. This new offspring M_2', is also exposed to the same sequence of input symbols, and its worth as a representation of the plant is similarly scored. If this discrepancy score is less than or equal to that of the parent, this offspring survives to become the new parent. If not, it is discarded, and a new offspring is generated. In this manner, nonregressive evolution proceeds to identify a succession of finite-state machines that are better representations of the plant, M.

At any point in time, the most recently evolved machine can be examined in order to determine that particular symbol which would serve as the most appropriate input for the plant, that is, for the present state of the machine, which transition will yield an output symbol contributing the least penalty in terms of the error cost matrix. After the most appropriate input has been identified, that input symbol is used to drive the plant. If a number of transitions are found to be equally suitable, then an arbitrary rule is applied to result in a selection. If additional sophistication appears justified, that transition is selected which leads to a state that offers the possibility of a wider selection of output symbols or of the particular output symbol that will next be desired, if that symbol is known.

At the same time, the best evolved representation for the plant is used for the generation of offspring that are successively evaluated over the available recall. Thus, the evolution continues in fast-time in preparation for the next selection of an input symbol for the plant. Such control decisions can be made to take place periodically, aperiodically, or upon request. They may be made to occur whenever a specified degree of proper representation has been achieved, when some pre-chosen number of offspring have been evaluated, or when a particular number of generations have occured. Generally, of course, the longer the time interval between control decisions, the greater the expectation of success. Similarly, the greater the speed of the computer facility (increase in the number of evaluated offspring) or the larger the available memory (increase of their permitted size), the greater the control capability.

As indicated above, the evolutionary technique offers distinct versatility in terms of the goal the system is to fulfill. For example, the purpose of the system might be to achieve minimum tracking error in the sense of a magnitude-of-the-difference cost matrix, a squared error cost matrix, or an "all or none" cost matrix. In fact, the desired performance might be specified in terms of a different worth for the correct tracking of each symbol and for each specific type of error. In other words, the error cost matrix need not be symmetrical in any regard. Nor is there a need for the goal to be time invariant. In fact, the goal may be made a function of the degree of success of past performance of the system if this should be desired. Many of the other features of evolutionary programming described in Chapter 2 are also implied.

Several experiments were conducted in order to demonstrate the evolutionary control technique. An arbitrary five-state machine served as the initial plant,[1] and at each point in time the best three machines were retained. Single, double, and triple mutation were made with equal probability. The goal was expressed in terms of a magnitude-of-

Table 5.1

355705156767626155265227255635367342704465755742241042150715362571472015
372410415071536267214720140235131531407162371520137553071524152011577643
765567101053230464751651153066745021727475733053531631262002014256732344
615403743307101627524535632577220102450464134417043234406104462200747737
555567507434125114442344702121472404561636522622703445666402503466571126
213716770413421401071717710033401511431653621063564044161467523614002562
741534566023401511307440731145447352025145005667447247014234006013645050

Table 5.2

Present State	1	1	1	1	1	1	1	1
Input Symbol	7	6	5	4	3	2	1	0
Next State	1	1	1	1	1	1	1	1
Output Symbol	0	1	2	3	4	5	6	7

the-difference error cost matrix. A penalty for complexity equal to 0.05 multiplied by the number of states was imposed. The reference signal was the uniformly distributed sequence randomly generated on 8 symbols, as shown in Table 5.1. Fifty control decisions were made in each experiment.

In the first experiment, the plant was simulated by the one-state machine shown in Table 5.2. Each control decision was made on the basis of the last 40 symbols of the reference signal, and the five possible modes of mutation were made to be equally likely. The first control decision occurred after an evaluation of 11 offspring, the best of these being a three-state representation of the plant. The control decision

[1] This being the same initial machine used in all of the 8-symbol experiments already described.

effected by this machine proved to be correct, thereby justifying its retention as a basis for the next decision. Although the second control decision proved to be in error, this same evolved machine was again retained as being acceptable for use in making the next control decision; this was done because the output of the plant agreed with the output of the model. In other words, in situations where those output symbols associated with the output of the model do not contain the desired symbol, minimum error is used as the criterion to determine the input to the plant. If, subsequently, the output of the plant agrees with the output of the model, the evolved machine is retained as being acceptable. The third control decision was again in error, but the output of the plant did not agree with the output of the model. Therefore, an active search was made for a better representation of the plant. Within the evaluation of the next 10 offspring on this recall, a one-state machine was found, which was identical with the logic of the plant. As a result, all further control decisions were correct. Note that if this plant had been directly driven by the reference signal, all transduction would have been in error.

In the next experiment, the one-state plant was replaced by a three-state plant that had only 0, 1, 2, 3, and 4 as output symbols. (See Table 5.3 and Figure 5.1.) The length of the recall was increased to include the last 100 symbols because of the increased size of the plant. The first 15 control decisions were based on seven different five-state machines. The next 5 control decisions were based on two different six-state machines. These were followed by 12 control deci-

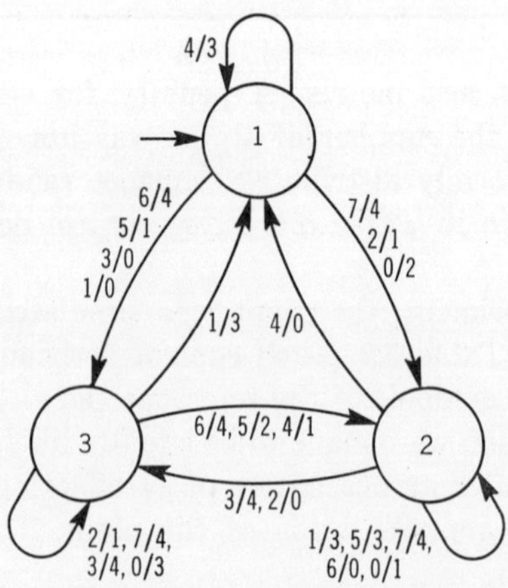

Fig. 5.1. Plant 2.

sions based on three different five-state machines. The last 18 decisions were made on the basis of the three-state machine shown in Table 5.4. This evolved machine has eight transitions that are identical with those of the plant.

Of the first 50 control decisions, 74 precent were correct (considering the symbol 4 in the controlled variable as being a direct response to the symbols 5, 6, or 7 in the reference signal). Learning was indicated, in that all of the last 14 decisions were correct. Of the 13 control decisions that were in error, 6 were within 1 symbol of being correct. For the sake of comparison, it is interesting to note that when this plant is directly driven by the reference signal, it is in error on 41

Table 5.3

Present State	1	1	1	1	1	1	1	1	2	2	2	2	2	2	2	2	3	3	3	3	3	3	3	3
Input Symbol	7	6	5	4	3	2	1	0	7	6	5	4	3	2	1	0	7	6	5	4	3	2	1	0
Next State	2	3	3	1	3	2	3	2	2	1	2	1	3	3	1	2	3	1	2	2	3	3	1	3
Output Symbol	4	4	1	3	0	1	0	2	4	0	3	0	4	0	3	1	4	4	2	1	4	1	3	3

Table 5.4

Present State	1	1	1	1	1	1	1	1	2	2	2	2	2	2	2	2	3	3	3	3	3	3	3	3
Input Symbol	7	6	5	4	3	2	1	0	7	6	5	4	3	2	1	0	7	6	5	4	3	2	1	0
Next State	2	2	3	1	2	3	1	2	3	3	3	3	1	1	1	3	2	1	2	3	3	3	1	3
Output Symbol	4	4	1	3	0	1	3	3	4	0	3	0	4	0	3	3	4	4	3	0	4	1	3	1

of the 50 symbols. Of these, 70.8 percent of the controlled variable symbols of the output sequence deviate from the reference signal by more than 1 symbol. In the limit, this plant would be in error 79.2 percent of the time if it were required to track a uniformly distributed random reference signal.

The third experiment was conducted with a five-state machine serving as the plant. (See Table 5.5 and Figure 5.2.) This machine was arbitrary, with the exception that every symbol appears as an output of each state, thus making it, in principle, completely controllable. In this experiment, the recall time was again lengthened to include the last 400 symbols so as to ensure adequate statistical confidence. To conserve computer time, the probability of the five modes of mutation was changed so that 95 percent of the time a next-state transition was modified, while 5 percent of the time the initial state was modified. Of the 50 control decisions made, 20 were correct and 16 were

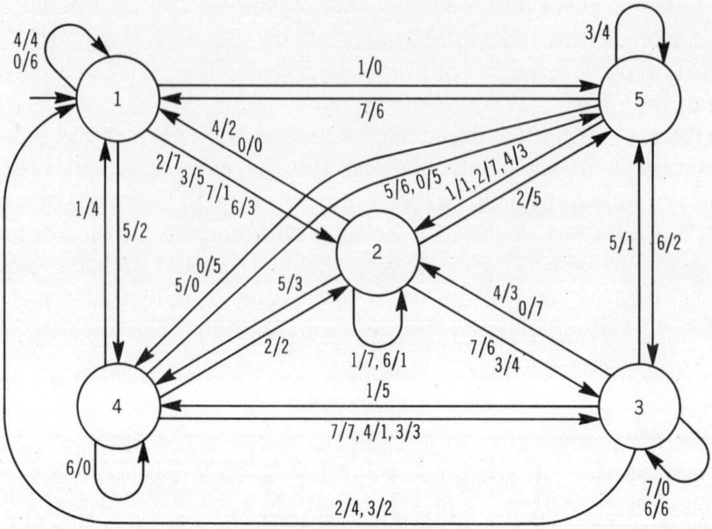

Fig. 5.2. Plant 3.

in error by only a single symbol. The last evolved machine shown in Table 5.6, had 23 of the 40 output symbols in common with the plant. In contrast, if this plant is directly driven by the reference signal, the output variable will track only 20 percent of the time, and will fall within a single symbol of the reference signal 24 percent of the time. In other words, system control through evolutionary programming provides an improvement of 100 percent in correct tracking of the random reference signal. This was accomplished with limited experience (450 symbols) during which none of the evolved models was a truly superior representation of the plant.

Table 5.5

Present State	1	1	1	1	1	1	1	1	2	2	2	2	2	2	2	2	3	3	3	3	3	3	3	3
Input Symbol	7	6	5	4	3	2	1	0	7	6	5	4	3	2	1	0	7	6	5	4	3	2	1	0
Next State	2	2	4	1	2	2	5	1	3	2	4	1	3	5	2	1	3	3	5	2	1	1	4	2
Output Symbol	1	3	2	4	5	7	0	6	6	1	3	2	4	5	7	0	0	6	1	3	2	4	5	7

Present State	4	4	4	4	4	4	4	4	5	5	5	5	5	5	5	5
Input Symbol	7	6	5	4	3	2	1	0	7	6	5	4	3	2	1	0
Next State	3	4	5	3	3	2	1	5	1	3	4	2	5	2	2	4
Output Symbol	7	0	6	1	3	2	4	5	6	2	0	3	4	7	1	5

Table 5.6

Present State	1	1	1	1	1	1	1	1	2	2	2	2	2	2	2	2	3	3	3	3	3	3	3	3
Input Symbol	7	6	5	4	3	2	1	0	7	6	5	4	3	2	1	0	7	6	5	4	3	2	1	0
Next State	3	1	5	2	5	3	2	1	2	2	1	2	3	4	4	4	1	1	1	2	4	3	1	4
Output Symbol	7	0	6	2	3	2	4	5	1	1	3	2	4	5	0	6	0	1	0	3	4	4	5	0

Present State	4	4	4	4	4	4	4	4	5	5	5	5	5	5	5	5
Input Symbol	7	6	5	4	3	2	1	0	7	6	5	4	3	2	1	0
Next State	5	5	1	2	3	3	5	2	3	3	3	2	5	2	3	5
Output Symbol	6	2	2	3	4	7	0	5	7	2	1	3	4	4	1	6

Traditionally, the greatest attention has been devoted to the study of linear plants and to the design of control systems that minimize the mean squared error. Although such systems are of interest, they represent only a tiny fraction of the real-world plants and control systems that deserve quantitative consideration. Evolutionary programming, as already described, can be used to enlarge this scope. Finite-state machines offer a convenient representation for any plant that can be adequately described in terms of a finite number of discriminably different internal states that react to input symbols with output symbols, both taken from finite alphabets. Even plants that change their behavior as a function of time can be controlled because of the iterative search to improve the representation. Certainly, the task of evolving a suitable representation becomes more difficult if the plant has internal disturbance, or if error is introduced in sensing the controlled variable; however, in principle, such difficulties do not preclude an evolutionary programming solution of the problem.

In keeping with convention, plants are usually described as being either linear or nonlinear. The need for such description is removed if the plant is expressed in the form of a finite-state machine. Viewed in this light, it becomes possible to associate with each plant a more meaningful descriptor: the degree of controllability. A completely controllable plant is one that can provide any of the possible output symbols for each of its state. A completely uncontrollable plant is one that has but a single output possible for each state. The domain connecting these extremes may be expressed in terms of the average ratio of the number of different output symbols per state to the size of the alphabet.

As long as evolutionary programming is restricted to the use of finite-state machines as the representation for the evolving organism, it is possible to encounter plants or environments that can never be completely expressed within the logic of a single organism. For example, the binary sequence 101100111000 . . . and the characteristic function of the prime numbers are sequences that cannot be perfectly described by a single finite-state machine. A single finite-state machine can predict without error only periodic environments. With occasional errors, it can describe an environment consisting of periodic sequences pieced together in quite arbitrary patterns. These sequences would be the different characteristic sequences of the machine corresponding to different initial state-input pairs. Errors would occur when it is necessary to switch from one sequence to another.

With regard to the above sequences, there exist rather simple finite-state machines that would predict them with an error rate asymptotically approaching 0. For the sequence 101100111000 . . . , since the length of the sequences of 0's and 1's is increasing, a machine with a trapped state (that is, a state which, once attained, cannot be left) whose output agrees with the input can approximate the sequence within any degree of accuracy. A similar machine in which the trapped state always puts out a 0 can describe the characteristic function of the prime numbers with a given accuracy for a sufficient number of predictions, since the frequency of occurrences of 1's in the first n terms is of order $1/\log n$.

As long as the alphabet of descriptor symbols is finite, a suitable coding can be found that will translate the environment into sequences of positive integers. The problem thus becomes one of computing numerical functions. Accepting Church's thesis (that the class of all numerical functions whose values can be computed by any means whatever coincides with the class of recursive functions), the recursive functions appear to be natural organisms for the evolutionary program. Such functions have been studied by various mathematicians. The development and terminology presented here is patterned after that of Davis.[2]

To introduce these functions, some preliminary concepts and definitions are required. If f and g are functions such that the range of g is contained in the domain of f, then the *composition* of f and g is the function whose value at a number x is given by $f(g(x))$. More generally, if f is a function of m variables and $g_1, g_2, \cdots, g_m$ are functions such that for each x $(g_1(x), \cdots, g_m(x))$ is in the domain of f, the composition of f and g_1 through g_m is the function whose value at x is given by $f(g_1(x), \cdots, g_m(x))$.[3]

[2] For proofs of quoted theorems and a more complete development, see M. Davis, *Computability and Unsolvability,* McGraw-Hill, New York, 1958.
[3] x itself may be an n-tuple, $n \geq 1$.

A function is *total* if it is defined for all values of the argument. If $f(y, x)$ is a total function, then the minimalization of f is the function $h(x)$ whose value for a given x is the least value of y if one exists, for which $f(y, x) = 0$. If no such y exists, the function h is not defined for that x; that is, h need not be a total function. For example, if $f(y, x) = x^2 + x + y^2 + 1$ then $h(x)$ is not defined for any x since the discriminant $1 - 4(y^2 + 1)$ is negative for all values of y and so the expression can never be 0 for any real values of x.

The function $x \dot{-} y$ is that function whose value is $x - y$ if $x \geq y$ and is 0 if $x < y$. If A is a set, then $C_A(x)$ is the function whose value is 1 if x is in A and 0 otherwise. C_A is termed the *characteristic function* of A.

A *partial recursive function* is any function that can be obtained by a finite number of applications of the processes of composition and minimalization, beginning with the following functions:

1. $S(x) = x + 1$
2. $U_i^n(x_1, \cdots x_n) = x_i$ for each n and i such that $1 \leq i \leq n$
3. $x + y$
4. $x \dot{-} y$
5. xy

If a partial recursive function is total then it is called a *recursive function*. If for some set A the characteristic function C_A is added to the above list, the class of functions obtained from the extended list is called the class of *A-partial recursive functions*. Similarly, any A-partial recursive function that is total is termed an *A-recursive function*. Thus, for any set A, each partial recursive (recursive) function is A-partial recursive (A-recursive). Moreover, since $C_\phi(x) = U_1^1(x) \dot{-} U_1^1(x)$ is the characteristic function of the null set ϕ, recursive is equivalent to ϕ-recursive.

Another closely related class of functions is the class of primitive recursive functions. This class includes all the numerical functions that are ordinarily encountered. Like recursive functions, they enjoy the property of being total and, in addition, have the following property: There is a binary recursive function $f(y, x)$ such that for any singularly primitive recursive function $g(x)$ there is an integer n such that $g(x) = f(n, x)$. Hence, the singularly primitive recursive functions are countable, and f provides an ordering for them. It is not true that all singularly recursive functions can be so derived from a single binary recursive function. To see this, let $f(y, x)$ be any binary recursive function: it then follows that $g(x) = f(x, x) + 1$ is a recursive function which is different from $g_y(x) = f(y, x)$ for each y.

To introduce the concept of primitive recursive function, it is necessary to define the operation of primitive recursion. If $f(x)$ and $g(z, y, x)$ are total functions then the function $h(z, x)$ where[4]

$$h(0, x) = f(x)$$
$$h(z + 1, x) = g(z, h(z, x), x)$$

is obtained from f and g by the operation of *primitive recursion.*

A function is *primitive recursive* if it can be obtained by a finite number of applications of the operations of composition and primitive recursion beginning with the following functions:

1. $S(x) = x + 1$
2. $N(x) = 0$
3. $U_i^n(x_1, \cdots, x_n) = x_i$ for arbitrary n and $1 \leq i \leq n$.

If the function C_A is added to the list, the resulting class is termed the *A-primitive recursive functions.* Further, by the same proofs as before, the class of primitive recursive functions is the same as the class of ϕ-primitive recursive functions and, for each set A, is contained in the class of A-primitive recursive functions.

The functions $x + y$, xy, and $x \dot{-} y$ are easily obtained by primitive recursion from the above functions and thus are primitive recursive. Moreover, suppose a function, h, can be obtained from two total functions, f and g, by primitive recursion. Then if f and g are A-recursive (recursive) so is h. Hence, the class of A-primitive recursive (primitive recursive) functions is contained in the class of A-recursive (recursive) functions. Further, the class of functions that are obtained from finite-state machines is contained in the class of primitive recursive functions.[5]

Another class of functions is the class of partially computable (computable) functions. A function is partially computable if there is a Turing machine that computes the function. If the function is total, then it is called computable. Further, the class of partially recursive functions is equivalent to the class of partially computable functions. Hence, the class of recursive functions and the class of computable functions are identical. By adding certain commands to the Turing machine that are effectively branching commands, depending on whether or not the symbol in the machine when the command is used belongs to a set A, the class of *A-partially computable functions* is obtained. This is equivalent to the class of A-partially recursive functions. Restriction to total functions gives the A-computable functions that are then identical with the A-recursive functions.

[4] x may be an n-tuple but z and y are single variables.
[5] Proof of this is included in Appendix D.

When these results are tied up with the previously cited ones, it becomes clear that there exists a Turing machine that computes the binary function $f(y, x)$ with the property that for every primitive recursive function $g(x)$ there is an integer n such that $g(x) = f(n, x)$. Hence, by either a random or deterministic selection of the integers n, it is possible to obtain the class of primitive functions by this Turing machine. Two problems arise. One is the practicability of simulating or constructing such a machine. The other is that even if such a machine were available, a suitable search scheme would have to be devised. While in principle an exhaustive search would eventually find any primitive recursive function, it may well be impractical. However, just the existence of such a machine is an exciting concept.

It is interesting to note that both of the previously mentioned sequences, that is, the characteristic sequence of the prime numbers and the sequence 101100111000 . . . , can be expressed exactly by primitive recursive functions. The existence of a Turing machine for the prime number sequence is given by Davis.[6] A machine for the sequence 101100111000 . . . is given in the Appendix D.

A Turing machine can be represented[7] as a finite set of quadruples having one of the following forms:

1. $(q_iS_jS_kq_m)$
2. $(q_iS_jRq_m)$
3. $(q_iS_jLq_m)$
4. $(q_iS_jq_kq_m)$

To interpret these, consider the machine as having a tape, assumed to be unlimited in both directions with one position being scanned. A quadruple of the form (1) specifies that if the machine is in state q_i and symbol S_j is on the position being scanned, then S_k is written in place of S_j and the machine goes to state q_m. Form (2) specifies that if the machine is in state q_i and S_j is the scanned symbol, the tape is moved right one position and the machine enters state q_m. Form (3) is similar except the tape is moved to the left. Form (4) is associated only with those machines which compute A-computable functions. If a set A is assumed specified, then if S_j is in A the machine enters state q_k, otherwise it enters state q_m. For the start of any computation, the machine is always assumed to be in a particular state called the start state with all but a finite number of tape positions being blank. The computation ends if and when the machine enters a state q_i with S_j

[6] Davis, *ibid.*

[7] This representation follows that of Davis; other forms are possible. In Appendix D a slightly different representation using quintuples is used.

as the scanned symbol where the machine does not have a quadruple beginning with q_iS_j. The result is then read directly from the tape. If the machine never enters such a state, the computation would go on forever; that is, the machine cannot compute the functional value for the given tape input, or equivalently, the function represented by the machine is not defined for that number.

Since such a representation is easily simulated in a computer, it is possible to replace the finite-state machines in the evolutionary program by Turing machines. While this would increase the domain of environments that could be perfectly expressed by the evolutionary program, it introduces a problem for which there can be no complete solution. This is called the "halting problem" for Turing machines. That is, can a technique be devised which would determine if a Turing machine will complete the calculation for a given number? Unfortunately, it has been shown that it is impossible to devise such a scheme, one that will work for all machines. It is an open question for what subclasses such schemes exist.

A possible move towards making the program feasible might be to put an upper limit on the number of steps for a single computation and reject any machine that exceeded this limit. Naturally, this would automatically preclude those recursive machines for which some computation would involve more steps than the prescribed maximum, and might possibly reject the machine being sought. The cost of this constraint is difficult to evaluate.[8]

In many cases, perfect decision-making is an enviable but unrealistic goal. The great majority of environments encountered can be suitably expressed in the form of the logic of the finite machine in spite of the fact that perfect decision-making may not be possible. Only with strong justification in terms of an expected class of environments would there be a justification for raising the level of generality of the organism. Versatility, the keystone of intellect, places greatest value on the use of the most general type of organism; however, the range of real-world experience provides an expectancy for encountering different classes of environment. This information can be used, in the interest of economy, to restrict the generality and, thereby, the versatility to within a suitable domain.

COMPETITIVE GOAL-SEEKING

To some extent, all decision-making entities are themselves affected by the response of the real world. That is to say, in principle at least,

[8] Appendix E provides additional discussion on this subject.

it is impossible to react to the observed environment without affecting it, at least to some degree. In the previous discussion, it was tacitly assumed that the influence of the decision-maker upon the environment is negligible. In other words, the communication link between the environment and the observer was considered to be essentially unilateral. At this point, it is of interest to consider the more general situation of bilateral communication wherein the decision-making entity both affects and is affected by its environment.

In his classic work, von Neumann considered the problem of devising optimal strategy in competitive gaming. He provided a solution for the player who desires to gain maximum payoff at minimum risk, finds himself in a zero-sum situation, and anticipates an unending game. Although considerable effort has been devoted to expanding the domain of applicability of game theory, it remains true that most real-world competitive situations present problems that fall well outside the very special class which are presently amenable to solution. It is suggested that evolutionary programming offers a new and different means for devising strategy over an extremely wide domain of competitive situations.

As a natural extension of the previous discussion, consider the situation wherein an organism exists within an external environment. This organism is equipped with sensory and motor capability, a memory, and an ability to process the sequence of symbols that are sensed in order to determine some desired response in the light of the given goal. The organism may be limited in terms of its ability to translate the result of such decisions into a corresponding action. That is to say, the physical properties of the motor capability may be viewed as forming a "plant" that serves to translate the results of decision into a controlled variable that is the organism's actual response to its environment. The worth of each response is indirectly measured by the organism in terms of the payoff associated with the future symbols it will sense. The measure of its goal-seeking ability is in terms of the cumulative payoff over the sequence of the sensed symbols thus far experienced.

The fundamental problem of the organism is identical with that of system control where it is desired to have an unknown and possibly limited plant track a reference signal, the cost of errors being expressed in terms of some given goal. As indicated earlier, this problem can be addressed through evolutionary programming.

A more difficult situation occurs if the plant can change its logical structure. Under such circumstances, it is important that there be a sufficient sample of input and output symbols in order to permit the construction of an adequate model of the unknown logic. It is difficult to estimate the size of such a sample, in that it greatly depends upon the number of symbols in the alphabet and on the degree of change in the logic. And yet, in principle, the evolutionary approach remains a valid approach to the problem, in that it affords a means for utilizing available information to the limit of the computation capability.

To generate the proper output symbol that is intended to influence the external plant, the organism must generate a model of its own plant. Greater difficulties arise if the sensory capability of the organism is so noisy[9] that a random disturbance is imposed upon the sensing of its own response. Such noise may be expected to degrade the capability of the organism to effectively model its own plant, and thus to optimize its goal-seeking performance. Here, again, though evolutionary programming appears to be a valid approach, the required sample size is also increased.

Even greater difficulty is introduced if the organism cannot directly sense its own response; that is, if the environment intercedes between the controlled variable and the sensed variable. Hopefully, the environment offers some relatively simple logic connecting these, a logic that changes little with time. Placing the organism in a noisy environment will handicap its ability to erect a model of the composite of its plant and the logic of environment in order to proceed further with its goal-seeking as measured on the sensed variable. Real-world situations usually include just such difficulties. The environment might contain other organisms that in turn respond to the activity of the subject organism. If the sensory capability of the subject organism is severely restricted, it may view the composite of such response in terms of a single descriptor at each point in time. Its task is then to form an adequate model of the entire environment as seen in toto. If however, the organism has sufficient sensory resolution, it may proceed to model each of the various "aspects" of its environment separately, and in this manner may choose a more suitable course in the light of its goal.[10]

It might be that these other organisms are also goal-seeking entities and modify their behavior as a function of the degree of success of the subject organism. Their performance then, might enhance or degrade the situation of the subject organism, but, in any case, its competition exists only on an absolute scale. Such an organism is engaged

[9] That is, the sensing of the internal stimulus to its own plant and to its own response can be in error.

[10] For a more detailed discussion of this point, see L. J. Fogel, *Biotechnology: Concepts and Applications,* Prentice-Hall, Englewood, N. J., 1963, Chap. 11.

in competition only at the lowest level; it cannot make use of devices such as bluffing, deception, collusion, and others that might serve to enhance its position, both on an absolute and on a relative basis. If its own behavior happens to benefit some other organism, there exists natural coalition, but such opportune activity is inadvertent.

For the organism to rise to a higher level of behavior, it would have to have an internal ability to separately sense its own stimuli and response in order to construct models of itself (these being distinct from the models of itself that may be constructed as reflections of its own behavior as seen in terms of its response of the environment). That is to say, a portion of the organism would have to be set aside for the purpose of representing the "self" in terms of some given goal (which, incidentally, may be different from the goal of the external performance of the organism). Note that the separation of such an information processing unit denies the organism the ability to model itself completely; this, in the sense that "the cat can never taste its own tongue." It is, of course, possible to set aside a section of this modeling capability for modeling the remainder; however, such successive sectioning soon finds an end in terms of limited utility under inadequate resolution.

But in order for such a model of self to be useful at any level, it must be compared with a model derived as a reflection of the environment. The comparison of these different kinds of models provides an ability for the organism to "describe itself" to its external environment. This is an expression of self-awareness, a basis for conscious behavior and communication at a higher level than "conditioned response."

In many situations, it is wise to assume that the environment contains at least one intelligent adversary, an entity which will modify its transduction so as to inhibit, if not prohibit, its control. Again, evolutionary programming is an appropriate way to proceed, except that here care must be taken to avoid the use of "inner loops"[11] that might ordinarily have been adopted in order to increase the efficiency of evolution. There is always the danger that the plant may model the "controller," at least to the extent that it can detect such regularities and, as a result, adopt an obverse policy. In general, greater efficiency in effecting control is gained at the cost of a loss in security.

Surprisingly enough, the additional "complication" of a bilateral communication channel may offer a distinct advantage to the decision-maker. With an ability to affect the behavior of the environment

[11] Internal feedback procedures.

comes the possibility to "interrogate" the environment by selecting stimuli and analyzing the observed response. Hopefully, there may appear some highly certain stimulus-response relationship so that the decision-maker may choose to avoid those actions that are expected to defeat its goal and take those other actions that have a significant likelihood of resulting in a benefit. In any case, the greater the number of "known" stimulus-response pairs of the environment, the greater the probability of choosing sequences of action that will lead to a more favorable situation.

In general then, the competitive decision-maker should have an ability to construct representations for the logic that can be used to "explain" the observed response-behavior of the environment. On the basis of such a model, the environment might be made to behave more predictably and in a manner more favorable to the decision-making individual.

In the elementary situation, a decision-making entity finds itself within an unknown deterministic environment. It has at its disposal a bilateral communication channel that serves to portray the condition of the environment in terms of some finite alphabet of descriptors and to transfer symbols from the entity to the environment, thereby possibly affecting its future behavior. The number of descriptor symbols is generally at the disposal of the entity, and is best chosen not to exceed the resolution of the instrument that permits this information flow.

For the sake of simplicity, let the decision-maker affect its environment in one of two possible ways. A symbol is sent to the environment as a query after each descriptor symbol is received. The task of the decision-maker is to choose that sequence of symbols which will elicit a sequence of descriptor symbols that will allow the construction of a finite-state diagram able to describe the logic of the environment to the greatest extent possible, thus making the future behavior of the environment more nearly predictable. If there is some cost associated with exercise of the communication channel, it is of interest to determine the minimum length interrogative sequence of action symbols that will completely reveal the logical structure of the environment.

Suppose the decision-making entity initiates its investigation with knowledge of the number of states the environment may assume. Further, let it be assumed that the logic of the environment is strongly connected. (That is, there exists some sequence of symbols that will cause the eventual transition of the environment from any state into any other state.) This assumption is reasonable because states that

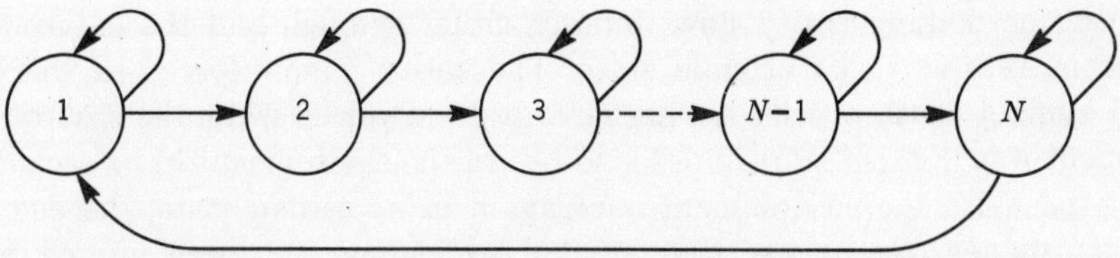

Fig. 5.3. The simplest strongly connected logic to discover.

are not strongly connected can only appear in the initial sequence of transitions. Such states can never recur, and thus are not of interest in predicting the long-term behavior of the environment.

With greatest fortune, a strongly connected diagram with N states and a 2-symbol alphabet can be fully diagnosed in as little as $2N - 1$ inquiries, since the last transition may be uniquely associated with this property. For example, Figure 5.3 indicates a diagram of N states that is constructed so that each inquiry reveals a return to the same state or a transition to the next state numbered sequentially. After $2N - 1$ transitions, there is only one possible transition that would make the diagram strongly connected, and this is a return to the original state.

It is also of interest to consider the logic that might require the longest sequence of symbols to reveal its logical structure. (Such a case might occur if an intelligent adversary were permitted to construct his logic as the sequence was received, but had not the ability to modify the logic once it became specified.) As shown in Figure 5.4, the first symbol is accepted by the environment without a change of state. It would be pointless to repeat this symbol against a deterministic environment, so the decision-maker goes on to choose the other symbol. For the finite-state machine of the environment to be strongly connected, this new symbol requires a change of state. The

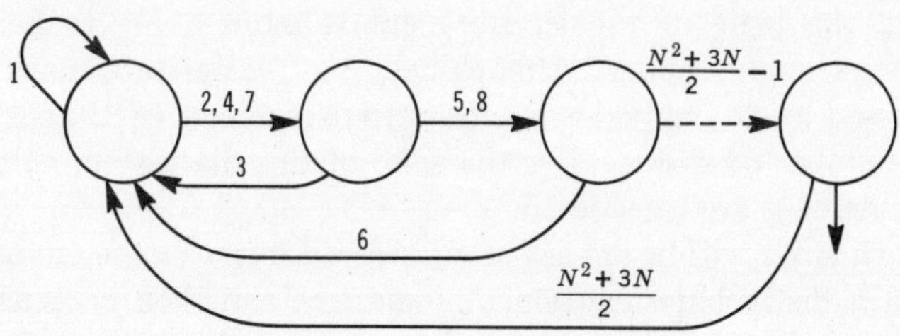

Fig. 5.4. The most difficult strongly connected logic to discover.

decision-making entity now imposes either symbol, and the environment returns to its original state. The entity is now forced to waste a symbol (both queries having already been voiced while the environment was in this state) in order to return the environment to its second state. With the environment once again in its second state, the decision-maker can impose that symbol which had not been offered to the environment while it was in that state. Again, for the logic to be strongly connected, the environment must transfer to a third state. The decision-maker follows with either symbol, and the environment transfers back into its original state. Now the entity must retrace two already-known transitions in order to return the environment to its third state and to make it able to ask the remaining question while the environment is in that state.

This same procedure can be extended to any number of states. After each new state is reached, there is a transition back to the original state so that a number of already-known paths must be retraced. Notice that there are N transitions into the original state, N transitions from the first state to the second state, one less transition from each state to the next higher state, and one transition from the last state to any state in the diagram, thus

$$\begin{aligned} N + N + (N-1) + (N-2) + \cdots + 1 &= N + N(N+1)/2 \\ &= (N^2 + 3N)/2 \end{aligned}$$

Hence, for any machine with N states, there exists a sequence of input symbols that will completely diagnose the machine where the length of this sequence, l, is between $(2N - 1)$ and $(N^2 + 3N)/2$.

It is of obvious worth to have some efficient technique for constructing a minimum-length sequence that, when offered to the environment, will provide a basis for determining its logic as long as it acts as a finite-state machine. Each symbol of such a sequence must be chosen in the light of the previous response from the environment. With knowledge of the number of states, it is possible to list the entire set of machines that are under consideration, but in the more usual situation, the logic of the environment is more likely to be one of some few strongly connected machines. Any finite-state machine can be expressed as an equivalent state-output machine (with, possibly, a greater number of states). For the sake of this discussion, only state-output machines are considered.

Let each state within the set of considered logics of the environment be given a distinct designation. A logic tree can then be constructed by starting with the state of "no-knowledge," this state bearing the composite designation of all of the states of the machines under consideration. The result of applying each of the possible input symbols to this state will be separately considered by noting the effect on that component of the composite state as seen within its original logic. Thus, for each input symbol, the state of no-knowledge will produce an output symbol and lead to a next state. New composite states can now be formed by combining those states which, for the same input symbol, yielded the same output symbol. In a similar manner, each of these composite states is examined in terms of each of the possible input symbols, resulting in the formation of new composite states. This procedure is continued until branches are found that terminate in states which are identified with states from only one of the considered machines. If a null set of states is reached, it may be discarded or used to identify that the environment does not follow the logic of any of the considered machines.

By definition, S_0 is this final set of "unique-knowledge" states that are identified with individual machines. S_1 includes the set of states which will transfer to unique-knowledge states through a single input symbol *and* the set S_0. The sets of states S_2, S_3, and so forth, are defined in a similar manner so that the addition of a single input symbol will lead only to states contained in the already identified preceding set of states. This procedure is continued until the original state of no-knowledge is included in a class. (As long as S_i does not contain the state of no-knowledge, S_{i+1} will properly contain S_i.) In general, a most efficient interrogation procedure as a function of each last response is that sequence of input symbols that causes a transition into the next, or still lower order, nested set of states regardless of the response, until S_0 is reached. A minimum-length interrogation sequence can thus be found for each of the considered machines by finding the shortest path from that machine to the state of no-knowledge, the sequence of symbols being those that lead along this path from the state of no-knowledge to the state identified with the considered machine. If the reply of the environment deviates from this path at any point, the considered machine can be discarded. If it is certain that the actual logic of the environment is one of the considered machines, then the final enquiry symbol will determine the proper machine. If, on the other hand, such certainty does not exist, the machine obtained by this procedure may not completely represent the proper logic of the environment, in view of the brevity of the interrogation sequence.

To illustrate, let the logic of the environment be presumed to follow either one of the two state-diagrams shown in Figure 5.5 As shown in Figure 5.6, the original state of no-knowledge is designated

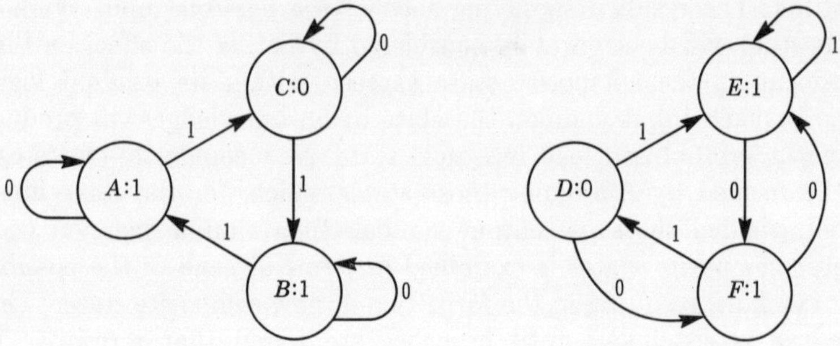

Fig. 5.5. A set of possible logics of the environment.

ABCDEF, this designation referring to all states of the considered machines. An input symbol 1 will take state *A* into state *C* and cause an output of 0; and 1 into *B* causes a transition to *A* with an output of 1, and so forth. Thus, of the considered machines, the two states that result from an input symbol 1 are *ABE* (which corresponds to an output symbol 1) and *CD* (which corresponds to an output symbol 0). In this example it happens that states of unique knowledge are those having a single designation letter. These states collectively form the set S_0. The set of states S_1 are those in S_0 and, in addition, those

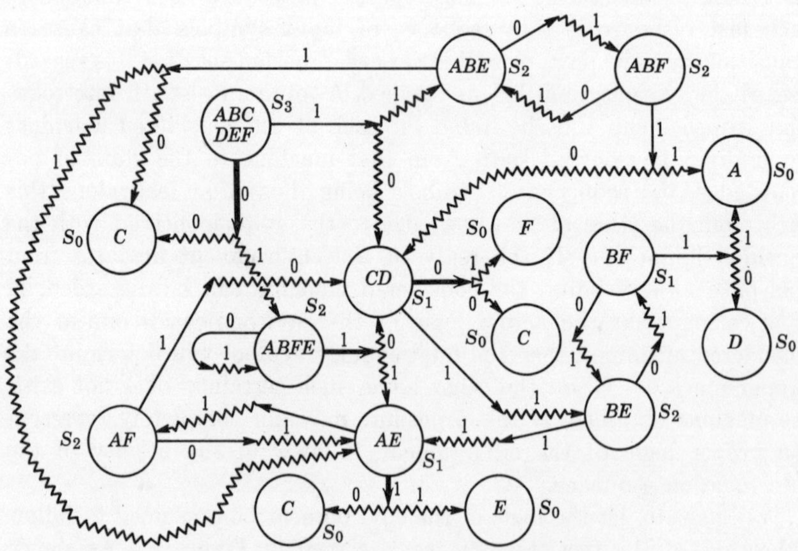

Fig. 5.6. The diagnostic tree of knowledge.

states which, for some input symbol, necessarily lead to a state of S_0. Higher-ordered sets are similarly defined.

The minimum-length interrogation sequence is now seen to be those input symbols that correspond to heavy arrows starting from the state of no-knowledge. An input of 0 might yield a response of 0. This would immediately terminate the inquiry in the knowledge that the environment obeys the logic of that machine which contains state *C*. A response of 1 would lead to the state *ABFE*. According to the logical tree, the next interrogation symbol is a 1. Here, a response of 1 will lead to a last interrogation of a 1, while a response of 0 will lead to a last interrogation of a 0. In any event, no more than three stimulus symbols need be applied in order to complete the diagnosis. The value of this technique becomes more and more evident as the number of considered machines is increased.

Other techniques for gaining "insight" into the logic of an interactive environment and for achieving effective utilization of this information may be found, but such problems are of significant difficulty. Evolutionary programming carried on simultaneously at various levels of abstraction[12] may provide an effective means for determining the logical structure of unknown and changeable interactive environments.

[12] Levels of abstraction are defined at an appropriate point in the next chapter.

CHAPTER SIX

Some Implications

THE SCIENTIFIC METHOD: AN EVOLUTIONARY PROCESS

The scientific method is an iterative process that facilitates the gaining of new knowledge about the nature of an observable environment. The evolutionary aspects of this process become apparent upon examining the process in some detail.

Initially, the investigator chooses to estimate the value of some unknown aspect of his environment. In the light of this goal, he collects whatever empirical evidence may be available from previous observations (removing, of course, those systematic errors known to exist). Although he recognizes the fact that these data are not without error, he presumes them to be true and proceeds to generate a model of the environment that is consistent with as much of these data as possible. This process of generalization is called *induction*. It constitutes the creative step of the scientific method. The models that result are given various names—hypotheses, conjectures, theories, representations, rules, language—depending upon the context.

A model is useful only if it provides a basis for an estimate of the specific unknown data point under consideration. Finding that specific estimate from the model is a deductive specialization. To facilitate this process, it is often worthwhile first to perform some deductive transformation of the model; that is, manipulating it into some more tractable form called a "general solution." Of course, the scientific method remains incomplete until there has been an attempt to verify or deny the validity of this estimate of the unknown data point through an independent observation of the environment. These logical steps are indicated in Figure 6.1. In brief, the scientific method consists of the induction of an appropriate model, the inductive inference of a specific unknown, and an independent measurement intended to verify or deny this inference; these steps are iteratively applied to the expanding base of known data points.[1]

With each measurement comes a benefit. Even if the estimate derived from the model is not verified, the effort is not in vain, for a new data point is now known. This enlarges the set of empirical evidence that may be assumed "true" for successive iterations. With an increased sample size, it should be possible for the investigator to create a better model, relative to some particular environment. As the knowledge of this domain grows there is greater likelihood that each successive estimate will be verified.

An additional and deeper benefit accrues whenever an estimate of the unknown is verified. Under such a circumstance, it becomes reasonable for the investigator to increase his belief in the value of that specific model as a valid descriptor of the environment; this being done in spite of the fact that he realizes that the truth of a model at some particular time offers no assurance that this same model will prove true at another time or, for that matter, for some other data point. Even a single verification should increase the relative worth of this specific model over those other models that could be generated upon a lesser foundation.

After each iteration the investigator may review those models that were true at the time they were exercised. It is natural for him to wonder whether this set of models might contain additional information concerning the environment; this information residing in their similarity or trend. With this in mind, the investigator may proceed

[1] The logical steps of the scientific method are described in more detail in Appendix E.

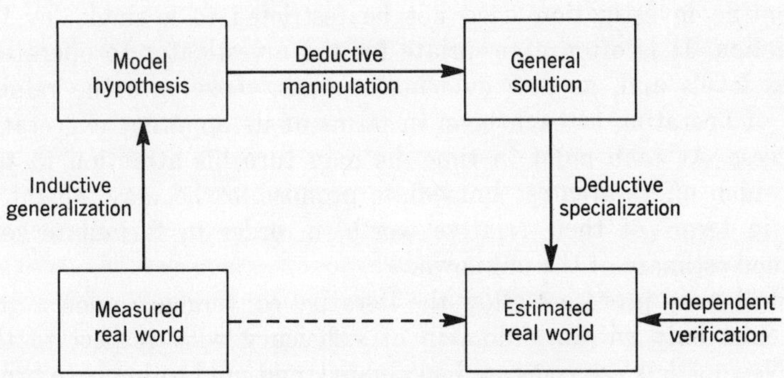

Fig. 6.1. The scientific method.

to exercise the scientific method at a higher level of abstraction. That is, he may choose to generate a model of those models that had previously proven true, a model consistent with as many of these previously successful models as possible. This new model (generally expressed within an alphabet of symbols different from those at the lower level of abstraction) can be specialized to yield another, as yet untested, model at the first level of abstraction. Hopefully, this model is more likely to be successful, in that it may embody some overall property of the environment that can only be found by viewing the range of experience.

In a similar manner, the investigator may rise to higher and higher levels of abstraction. Each new level is reached by forming a model of those models that have previously proven true at the last level of abstraction. Models of the directly observed data lie at the first level of abstraction. Models of the sequence of those models that have proven true lie at the second level, models of the sequence of these that have proven true lie at a third level of abstraction, and so forth. Note that models of a higher level of abstraction must be successively reduced into first-level models in order to be tested for their validity against the observable environment. In general, fewer and fewer models comprise the set of available evidence at higher and higher levels of abstraction. As a result, the models that are generated are less likely to be verified. Increasing the level of abstraction provides the advantage of an "overview," but this benefit is reduced by the increased probability of error. The additional effort associated with operating at a higher level of abstraction is balanced by the greater payoff associated with success. Verification of a high-level model may obviate costly step-by-step exploration of the intervening portion of the environment.

Scientific investigation need not be restricted to a single level of abstraction. It is often appropriate for the investigator to operate at various levels and, as new evidence accrues, reevaluate the relative worth of operating at each level in terms of its apparent expectation of success. At each point in time, he may turn his attention to that level which offers greatest immediate promise, or he may weight all levels in terms of their relative worth, in order to formulate some combined estimate of the unknown.

Evolutionary processes allow the iterative convergence upon a goal. They exist only within a domain of sufficiency with respect to that goal. New models are generated, evaluated, and used to improve previously held models only when their evaluation indicates that their retention is likely to help meet the challenge of the environment. In biological terms, each newly created organism, generated by previously "successful" creatures, is matched against its environment. Only those that exceed minimum requirements for survival remain to supercede their progenitors. Such successive selection contributes to the survival of the species. Each living creature may be viewed as a tentative model of some significant aspects of its environment. The domain of this evolutionary exploration is dependent upon the nature of the reproductive process and the diversity contributed through natural mutation.

In a similar manner, social groups are formed as a result of common needs in the continual struggle against the antagonistic environment. They are organized of individuals who have thus far proven to be successful at the personal level. Their structure reflects the character of these individuals and of the logic of their environment.[2] The family includes the individual, the community includes the family, the state includes the community, and so forth. At each level, the structure evolves so as to reflect the nature of demonstrated success at all lower levels of abstraction with respect to the chosen goal and the properties of the encountered environment.

The efficiency of pure trial-and-error exploration is sharply reduced with an increase in the dimensionality of the domain being explored. As long as the investigator is interested only in a single aspect of his environment, random exploration may prove to be worthwhile, but as soon as he attempts to map a domain of more practical interest he encounters so many possibilities that only carefully guided trial-and-error exploration is likely to prove profitable. The advantage to be gained from employing the scientific method becomes self-evident. In man's initial exploration of the unknown, the scientific method would have been a luxury; however, with the increased scope and depth of his inquiry, use of the scientific method becomes an absolute necessity.

The scientific method consists of induction, inductive inference, followed by independent verification. Hypotheses, generated so that they "cover" the available evidence and some additional data points, are individually evaluated in terms of the validity of their inference. Those that prove worthy are modified, extended, or combined to form new hypotheses that carry on a "heredity of reasonableness." As the hypotheses correspond more and more closely with the logic of the environment they provide an "understanding" that is demonstrated in terms of improved goal-seeking behavior in the face of that environment.

[2] All environments may be viewed as antagonistic, in that the sheer passage of time tends to disorganize and thus threaten the survival of the decision-maker.

The correspondence between natural evolution and the scientific method is obvious. Individual organisms in nature serve as hypotheses concerning the logical properties of their environment. Their behavior is then an inductive inference concerning some as yet unknown aspects of their environment. Their survival is a demonstration of their suitability. Their offspring include the heredity of reasonableness as well as additional information resulting from mutation and recombination.

The evolutionary technique described above is then a realization of the scientific method in which the hypotheses are restricted only in the sense that they be finite-state machines. The creation of successive hypotheses requires the introduction of randomness as well as the prevailing logic of inheritance. The versatility that is essential to intellect is a natural product of such iteration. In essence, the scientific method is an essential part of nature. It is no wonder, then, that its overt exercise has provided mankind with distinct benefits and now permits even its own automation through the artificial evolution of automata.

THE AUTOMATION OF INTELLECT

Viewed in this light, the scientific method was not invented, it was discovered. It existed long before man; in fact, it gave rise to man. Natural evolution can be looked upon as a realization of the scientific method.

As indicated, some of the environmental aspects of evolution can indeed be simulated. In essence, such simulation constitutes a mechanization of the scientific method. The successive generation of models . . . the process of induction . . . has been reduced to a routine procedure. If "creativity" and "imagination" are requisite attributes of this process, then these, too, have been realized. But, according to the definition offered at the beginning of this discussion, intelligence is "the ability of any decision-making entity to achieve a degree of success in seeking a wide variety of goals under a wide variety of environments." The scientific method provides an appropriate foundation for just such decision-making. In principle, the nature of the goal and the nature of the environment are in no way restricted. In a very real sense, mechanization of the scientific method provides a basis for the realization of artificial intelligence. In 1950, Turing said: "We may hope that machines will eventually compete with men in all purely intellectual fields."[8]

[8] "Can A Machine Think?", by A. M. Turing, in *The World of Mathematics*, Vol. 4, edited by J. R. Newman, Simon and Schuster, New York, 1956, p. 2122.

The first phase of computer technology was devoted to the development of equipment that would carry out a large number of simple logical operations quickly and reliably. The second phase of computer technology was devoted to the development of languages that permit the detailed instruction of these equipment in an efficient manner. Computer technology is now entering a new phase, one in which it will no longer be necessary to specify exactly how the problem is to be solved. Instead, it will only be necessary to provide an exact statement of the problem in terms of a "goal" and the "allowable expenditure," in order to allow the evolution of a best program by the available computation facility. Solution of the problem may include a statement of the discovered algorithm; or, if requested, the entire history of the evaluated algorithms. The old saw "the computer never knows more than the programmer" is simply no longer true.

It would, of course, be inappropriate to utilize evolutionary programming for the solution of problems that have already been solved. There is always an extra price to pay for versatility. That additional fee should be offered only in situations that justify the search for a new method of solution—a new logic for decision-making. At this point, the discussion should be turned to economic details; unfortunately, however, at present little is known about such details. The programs that were used for the conduct of the experiments described earlier were inefficient in many respects. They were written to demonstrate feasibility rather than efficiency. Evolutionary programming should prove applicable to *any* digital computer of a reasonable size. The particularities of the program and the related cost of operation may be expected to depend strongly upon the characteristics of the equipment and the available programming language. Further studies in this regard are worthwhile endeavors.

At this point, it is natural to inquire, "Is there an upper bound to intellect?" Beyond the obvious constraints imposed by the sensory and motor capabilities, are there any fundamental aspects of information processing which limit the level of intelligence of any given entity? In terms of evolutionary programming, there are, of course, physical properties of the equipment which limit what can be achieved. The speed of computation limits the number of offspring that can be evaluated within the time available for decision. The size of the memory limits the specificity of the individual models as well as the number of models that can be retained at each point in time. It is certainly possible to increase the available memory capacity, but this increase is usually accompanied by a considerable increase in information access and retrieval time. This, in turn, again limits the number of evaluated offspring. Clearly, these factors are strongly interdependent, the best compromise being a function of the particular hardware to be used for the data processing.

Future computers may be faster than presently available equipment. They may permit parallel data processing with obvious advantages. Such sophisticated machines may take the form of geographically separate data-processing elements that are tied together through a central control station. Such a computer may well operate in a decentralized manner so that local data processing may be carried out as a separate function. The central control would be called upon only for problems that require additional capability or for problems which affect the entire system "under its cognizance." A new physical limitation is introduced in such dispersed decentralized computer systems, in that the time of information transfer from element to element may become a significant factor. The capability of the entire decision-making entity may be restricted by the sheer point-to-point limitations of the physical embodiment. A number of currently extinct creatures appear to have met their demise as a result of just such internal "transmission delays." Clark suggests that the limitation of the intelligence of a galactic-sized decision-making entity would be limited by the speed of light which imposes unavoidable transmission delays.[4] It is easy to envisage other limitations caused by the size of the decision-making entity, for instance, being too large may be as bad as being too small.

It is of greater interest to consider some of the logical limitations on the intelligence of any goal-seeking entity. As indicated above, intelligent behavior may be expected only if the decision-making entity possesses some suitable model of its environment; the degree of suitability being measured in terms of the level of correspondence between the model and its real world counterpart. Therefore, intelligence is limited if the model is too coarse, thus failing to offer a sufficiently specific description of the details of the environment; or if the model is an incomplete portrayal of the relationships among the elements.

These points are worthy of clarification. The sensory capability of the entity provides some maximum resolution both in time and in the number of available descriptor symbols. The greater this resolution, the greater the available information base. The speed of response and the specificity of the modeling process further constrain the representation of the environment. Models can only be developed within

[4] A. C. Clark, "From Lilliput to Brobdingnag," *Playboy Magazine* (Vol. 9, No. 6), June, 1962, pp. 102, 116–119.

the available alphabet of relations. Inadequacy of this set of descriptors limits the accuracy of the resulting representation of the environment. For example, restriction of the model to a specific class of finite-state machines forbids perfect representation of environments that fall outside this domain. If the model can take the form of any finite-state machine, it may still prove inadequate for perfect representation of some primitive recursive environments. Similarly, the availability of primitive recursive machines would still forbid perfect representation of more general recursive environments. Restriction of the generality of the modeling process must be justified in terms of the adequacy of representation over the anticipated class of environments. Fundamental research is continuing in an effort to find a more complete specification for the relationships that are found to exist in the real world.[5]

Then too, the intelligence of an entity may be restricted in terms of the parameters of its modeling process. For each situation, there are most suitable values for the multiplicity of mutation, the weighting of recall, the available levels of abstraction, and so forth. In addition, some selection must be made of an operating point between the extremes of efficiency and security.[6] If the environment is presumed to be almost independent, increase of efficiency may well be justified. If, however, the environment might prove to be an intelligent adversary it may be more appropriate to forsake efficiency for the sake of security. The most intelligent player need not be the most efficient player.

Only goal-seeking entities can be intelligent. Thus far, the goals have been described in terms of the relative worth (or penalty) of each possible outcome of the decisions to be made. At a more fundamental level, the goal of an entity is a description of some desired future status of the system which contains that entity and its environment. In the experiments reported above, the goal was specified to the entity by edict. In this sense, the *desire* was that of the experimenter and not that of the entity. Several significant problems are raised. What is meant by the desire of an inanimate entity? How can an entity construct a model of a preferred future status? How can qualitative statements of goal be translated into quantitative expressions suitable for exact computation?

[5] G. Gunther, "Cybernetic Ontology and Transjunctional Operations," in *Self-Organizing Systems,* Edited by M. C. Yovets, G. T. Jacobs and G. D. Goldstein, Spartan Book Company, Washington, D. C., 1962, pp. 313–392.

[6] A similar compromise occurs in aircraft design where an operating point must be selected in the range from extreme maneuverability to complete stability.

To be useful, the goal must be represented in some way within the entity. "Desirability" of any particular goal presumes the existence of a decision-making capability that can select a certain model from a set of possible models as being of greater value. This "value judgment" is a reflection of some higher-level goal that, in turn, must be selected on the basis of a still higher goal. This infinite regress can be avoided by allowing the entity an ability to model itself. As indicated in the previous chapter, it should be possible to design a mechanism which would exhibit self-awareness. Such a mechanism could describe the essential features of its own information processing if so requested. With such a capability, the entity may exhibit a natural tendency toward self-preservation,[7] this being an extension of its physical inertia. Self-preservation is the highest goal in whatever complex hierarchy of subgoals the entity may hold since all subgoals are contingent upon survival. The set of possible futures is generated by combining the results of a variety of possible events. That future status is considered best which offers the greatest expectation for preserving the self as portrayed by the available model. Here again, the procedure is evolutionary.

The specificity of models at various levels may indeed be quite different. At higher levels the goal may be an expression of gross intent. The model of the self may be highly unspecific. How specific must this representation be in order to "qualify" as a valid goal or provide a satisfactory demonstration of self-awareness? Once again, these limitations are directly dependent upon the physical embodiment of the entity in terms of its capability for information processing and storage. Just as there are degrees of intelligence, there are degrees of self-awareness.

[7] Challenge might be raised to calling self-preservation the topmost goal of living creatures by pointing to suicide and martyrdom as acts that demonstrate the dominant desire for self-destruction. Except for cases of insanity, rational individuals justify their behavior to themselves in some "reasonable" manner. It is contended that the martyr sacrifices his life to preserve some principle with which he identifies himself. He recognizes that the ideals he stands for may best be preserved through sacrifice of that which he holds to be of less value to himself. Within his rationale, he pursues self-preservation, but the self he preserves does not correspond to the biological entity that is his body. In a similar manner, suicides occur when individuals imagine that this act will allow them to escape into a better situation, will punish some other individual, or will provide some other form of revenge. Here again, the individual acts so as to preserve his own identity rather than his physical being. Other situations which on the surface appear to be cases of self-sacrifice may also be viewed in a similar light.

If sufficient sensory-motor resolution exists, it may be possible for the entity to separately model individual aspects of its environment in terms of their observed behavior. It may become possible to generate a presumed goal for each observed entity and compare this goal to its own goal. Recognition of similarities and differences in goals provides a basis for social behavior. Any modification of the behavior of interacting entities establishes the existence of an organization that, in itself, constitutes a new entity. The stronger this interaction, the stronger the organization. Its goal reflects the goal of its primary constituents. In recognition of its existence, it may take on a specific embodiment. Its goal gains overt expression usually in an attempt to more nearly reflect the weighted objectives of all its members. The family, the company, the community, and the nation are examples of just such organizations.

The subgoals of this organization must be generated in the light of its goal and those of its members. With demonstration of successful stimulus-response behavior, the organization gains in strength and stature. It soon becomes important for each member to weigh more heavily those subgoals that support the goal of the organization to the greatest extent. Allegiance is established. The members can no longer afford to view themselves as being autonomous. The essential similarity between intelligent entities (if nothing more, their intellect) tends to make their goals similar. Thus, they engage in competition. Only a higher level of social allegiance can ameliorate the resulting antagonism.

Viewing other entities in terms of their presumed goal can be dangerous. Any finite sample of observed behavior can always be "explained" in terms of any one of an infinite number of goals (including no-goal, the behavior of the entity being completely random). The selection of one of these goals as a descriptor of the observed entity appears to remove uncertainty from the situation, but this may introduce a cost in terms of security. Often, the policy is to presume that the observed entity is of equal intelligence to that of the observer. The question becomes: "If I had acted in that manner, what would have been my goal?" This is a conservative policy, but there is always the chance that the observed entity may have greater intelligence and thus take advantage of this supposition as the basis for a costly interchange.

There is an even more fundamental danger in generating "presumed goals" in order to rationalize the observed behavior. Note that some orderliness can be found in any finite sample drawn from a perfectly random environment. This orderliness is a reflection of the observer's own characteristics. In the more general case, the environment is unknown and may be orderly to some extent. Is it possible for the observer to separately identify that portion of the observed orderliness which is a reflection of its own characteristics from that portion which is truly indigenous to the observed environment? The answer to this question hinges upon the ability of the observer to "know" itself. In order to contain the model of itself, some portion of the entity must be devoted to this purpose. This portion of the entity is not described within the model. It is, of course, possible to have a section of this portion set aside in order to model the remainder of the portion, but, here again, the representation cannot be complete. By analogy, "the cat can't taste its own tongue." This ultimate limit also places a limit on the amount of orderliness that can be identified as a reflection of the observer. Discrimination of the orderliness of the observed environment is further degraded since each stimulus of the entity affects the observer to some degree, thus modifying its "interpretation" of successive stimuli and its model of self.

Intelligence is to some extent self-limiting. As the intelligent decision-maker gradually gains control over its environment, the severity of the remaining task is reduced. Only new goals or a radical change of environment will exercise the full intellectual capacity. The greatest development of intellect requires its continual exercise against a suitably antagonistic environment.

CHAPTER SEVEN

Conclusion

At this point, it is appropriate to review the argument in perspective. Intelligence was defined in terms of the ability of any decision-making entity to achieve a degree of success in seeking a wide variety of goals under a wide variety of environments. Versatility was recognized to be the essential ingredient of intelligent behavior. After a brief survey of bionics and heuristic programming, attention was focused upon the possibility of simulating evolution as a more general means for creating artificial intelligence.

To make the discussion specific, intelligent behavior was viewed as a composite of an ability to predict the sensed environment, coupled with an ability to select a suitable response in the light of the prediction and the given goal. The problem of predicting each next symbol was then reduced to the problem of finding that finite-state machine which would have been best for achieving this goal had it existed during the available experience. This search is accomplished through a fast-time simulation of evolution in which finite-state machines play the role of the evolving organisms. Random mutation of an arbitrary machine yields offspring. These machines are driven by the available history and evaluated in terms of the given goal. That machine demonstrating greater worth is selected to serve as the new parent. This mutation and selection is iterated with real-time decisions being based at any point on the logic of the surviving machine. The efficiency of evolution can be improved in a number of ways, as for example, by introducing a cost-for-complexity weighting on each machine.

But the goal need not be restricted to the prediction of each next symbol. In fact, this same evolutionary procedure is suitable for seeking any well-defined goal within the constraints imposed by the allowable expenditure. Thus, the evaluation may take place in terms of the response behavior, prediction becoming an implicit intervening variable.

A number of prediction experiments were conducted against 2-symbol and 8-symbol environments in order to demonstrate the feasibility of evolutionary programming. In particular, cyclic environments were found to be readily predictable even when the "signal" was disturbed by increasing levels of noise. The predictor-machines were shown to characterize the signal in terms of their asymptotic cyclic behavior when driven by their own output. Stationary time series were also predicted with considerable success, especially in comparison to the ability of human subjects.

A variety of Markov processes were predicted with considerable success. In the case of an arbitrary first-order process the evolutionary program attained the same score as would an optimal predictor even without *a priori* knowledge that the environment was stationary, Markov, first-order, and generated by a particular transition matrix.

Experiments against nonstationary environments included an attempt to predict the primeness of each next symbol in the sequence of positive increasing integers. With an imposed penalty for complexity, a trade-off point was reached beyond which it was not "worthwhile" to predict other than 0. The use of a payoff in proportion to the rarity of each properly predicted symbol made the problem even more difficult. Of the first 100 prime numbers only 5 were missed; the behavior of the evolutionary program reflecting a gradual construction of the definition of prime number. Note that *all* of these experiments were conducted using the same evolutionary procedure.

Other experiments were conducted in order to demonstrate an ability to characterize an unknown transducer in terms of the evolved predictor-machines. The input variable and output variable of the transducer were combined to form the input variable to the evolutionary program. The predictor-machine that evolved at the end of the experiment was translated into a representation of the logic of the transducer (an arbitrary five-state, 8-symbol machine). This representation was found to be exact, except for a single extra state that would presumably disappear if the evolutionary experiment had been continued further.

Experiments were conducted to demonstrate that evolutionary programming can be used for diagnosis. At each point in time, it is required to estimate an unmeasured parameter on the basis of the set of concurrent parameters and the available history. The correctness of each decision is then made known. A fictitious sample of data was generated in order to remain within the 8-symbol capability of the available evolutionary program. The results of these experiments demonstrated a significant capability not only in terms of correct diagnosis, but also in terms of sensitivity analysis. That is, the experiments revealed the rank order importance of the sensed parameters in terms of their contribution to making the required decisions.

A number of experiments were conducted in order to study the capability of the evolutionary program to recognize and classify patterns. Pattern recognition was readily achieved, but it was immediately evident that pattern classification was not being accomplished in a manner similar to that of an average human observer. Modification of the goal to more nearly reflect that of an average observer was shown to yield similarity in the classification capability. Of course, if this goal were known, there would be no point to using the evolutionary procedure, since this knowledge might be used more directly in terms of a heuristic program. The value of evolutionary classification is realized when it is required to identify some underlying similarity among the patterns that is *not* evident to the human observer.

In another series of experiments, the problem was to control an unknown plant so that it would track an arbitrary sequence of symbols. The plant could be linear or nonlinear as long as it was representable by a finite-state machine. The evolutionary program was used to find successively better models of the given plant in terms of its input-output behavior. At each point in time, the best available models were examined in order to determine that input symbol which would cause the plant to respond in the desired manner. If the plant could not produce all of the output symbols, its behavior followed the desired "track," at least within the limits of its capability.

Various aspects of competitive goal-seeking were considered. Simulated evolution may offer a means for approaching the general situation in which the goals of the players may take arbitrary form and the players interact through a common environment. Such a player may model his opponent and in this manner attempt to achieve the desired control over the situation. All of the already-described experiments concerning detection, discrimination, diagnosis, pattern recognition and classification, system control, sensitivity analysis, identification of the goal, and competitive gaming were carried out using a basic evolutionary program. Certainly, this is a demonstration of versatility.

But there are even more important implications from gaining artificial intelligence through simulated evolution. The scientific method may be viewed as an evolutionary process in which a succession of models are generated and evaluated. The scientific method affords a versatile procedure for gaining new understanding of the nature of the observed environment. Therefore, simulation of the evolutionary process is tantamount to a mechanization of the scientific method. Induction—that process which was presumed to require "creativity" and "imagination"—has been reduced to routine procedure. Here is a technique that offers an approach to the automation of intellect, relieving the investigator so that he can turn his attention to higher-level problems of selecting suitable goals and identifying new parameters of the environment that are worthy of measurement.

As indicated, the scientific method may be operated at various levels of abstraction, each higher level operating on the environment created by those models of the lower level that were of demonstrated validity. At the same time, it should be possible to devise mechanisms that will generate their own goals by reference to their internally held model of self and the "desire" of self-preservation. Such an entity might exhibit self-awareness, in that it could describe essential features of its own information transduction if so requested.

With increased sensory-motor resolution comes an ability to separately model individual aspects of the environment, to recognize the existence of similar entities, to interact with these, and, thereby, to establish a social organization. This organization in turn assumes the role of a decision-making entity with respect to its environment. Indeed, it may exhibit the same kind of goal-seeking behavior that is characteristic of its members. Their choice of goals must then take place within the context of various levels of allegiance.

Memory is built on memory. All that is sensed is lost, except for those items that can be associated with past experience in terms of identical factors. All of these items are linked by the bond of the common value scale in terms of the demonstrated worth of previous elicited response. Thus, without a goal, memory is blind and even survival hinges upon pure chance.

And what are goals made of? They are made of the factors that contribute to self-preservation, the invariance of identity of the entity as it sees itself. Only those creatures that can effectively model themselves can alter their subgoals in support of their own survival. To survive, the self-image must be in close correspondence with the reality of themselves and their environment.

Here lies the chronicle of life. With this knowledge comes a new power to enhance memory, to manipulate goals, and to make better decisions. Of even greater significance is the possibility of creating inanimate mechanisms that possess these same qualities.

Epilogue

Human sensory and motor capabilities have been significantly extended through technology. Effort has been concentrated on the problem of allowing man to examine his surround. Instruments have been developed that permit him to measure with increased sensitivity and precision. At the same time, great effort has been devoted toward extending man's motor capabilities. Through technology, he has provided himself with machines that can manipulate his environment and amplify his own ability to move. Vehicles are, in a very real sense, an extension of man's motor capability.

Unfortunately, relatively little attention has been devoted to enhancing the decision-making process that translates sensed information into selected response. The recent advent of high-speed computing equipment makes it possible to relieve man of the drudgery of routine decision-making. Today, most of the responsibility for the transduction still rests with the man; however, even as he accepts the machine-provided answers without individually checking them, he delegates some responsibility to the machine. As problems become more complex and computers become more sophisticated, an increasing portion of the total responsibility will be given to the machine.

Recognition that the scientific method is an evolutionary process that *can* be simulated opens the possibility of automating higher levels of intellect. Such evolutionary programming goes beyond the benefit provided by the sheer speed and accuracy of modern computing equipment. Here is a way to discover new techniques for goal-seeking even without human intervention. In its modeling of the environment, the machine exercises its own "creativity" and "imagination." Surely this is artificial intelligence. With this demonstrated, it becomes possible to seriously consider programs that would provide the machines with self-awareness and, ultimately, with an ability to select its own goals.

The automation of intellect poses two dangers to mankind. Removal of the responsibility for decision-making presents a threat to the continued existence of the ability of man to make such decisions. Only a succession of significant challenges can prevent the atrophy of human intellect.

The second danger is more subtle. The interaction of goal-seeking entities produces an organization that itself responds to stimuli so as to reflect the goals of its members. With formalized embodiment, such an organization can, in a subtle manner, usurp the prerogative of its members to choose their own goals. In the interest of its own self-preservation, their goals are forced to become aligned with the goal of the organization. In the limit, the result is an organization dedicated only to its own perpetuation; the individual goals of its members are supported only as these happen to coincide with *its* goal. Totalitarian nations are a case in point.

In the interest of short-term security, nations that compete with totalitarian powers for survival are pressed to adopt just such a centralized structure in order to achieve greater internal efficiency, flexibility of policy, and speed of response. But the goal of nations should be of longer term. We must attempt to look beyond the immediate threat.

The greatest benefit to all mankind can only be derived through freedom of the individual to select and pursue his own goals, so long as this activity does not infringe upon the rights of other individuals. Only through a decentralized decision-making organization can this status be maintained. For such a nation to survive, the "man on the street" must assume responsibility for learning about his environment, about himself, and about what can be achieved through artificial intelligence. Technology benefits only those who prepare for it.

APPENDIX A

Prediction Restricted to Cyclic Models of the Environment

For the sake of simplicity, the environment is viewed as being in one of two possible conditions at each point in time. It is of interest to examine a set of cyclic predictive models of increasing complexity that an observer might make to remain consistent with the sequence of observed data within his limited memory.

Prior to observing the environment in terms of a set of descriptor symbols, the individual cannot be expected to make any worthwhile prediction. Without exercising his observation capability, the decision-maker has no way of identifying the particular kind of descriptor symbols he can use.[1] Upon observing the first descriptor symbol in time, the simplest and most natural assumption is that only this symbol exists. Therefore, the information-processing individual predicts that this same symbol will appear again at the next observation of the environment. In a similar manner, if the first observation of the environment had revealed a sequence of two identical descriptor symbols, the decision-maker would have predicted the same symbol to occur a third time. In fact, if all the experience consists only of some finite sequence of the same descriptor symbol, the observer can reasonably presume that the environment is invariant and, on this basis, predict that this same symbol will occur again.

Suppose, however, that the first view of the environment had re-

[1] Let it be assumed that the decision-maker has prior instruction that an environment exists and has, in fact, existed long before his initial observation.

vealed an alternation of the descriptor symbols: The decision-maker may then choose to predict that the most recent symbol will reappear, on the grounds that the environment is inertial. This is a minimum relaxation of the assumed constancy. On the other hand, the observer of such an alternating pair of symbols may choose to operate within a new alphabet of symbols comprised of successive couplets of the previously observed descriptor symbols, this being tantamount to the presumption of a cyclic environment. With this transliteration, it becomes possible to return to the assumption that the environment is invariant, and, on this basis, to predict the first of the two individual symbols to appear as the next observation (in that the couplet is expected to reappear). (In this argument, it is presumed that the decision-maker accepts his observations as being errorless and, therefore, will not assume a constancy of the environment in terms of either of the observed symbols.)

In the situation where the only observation of the environment reveals a triplet of alternating individual descriptor symbols, prediction of a reappearance of the last symbol can be made on the grounds of either its recency (the assumption of an inertial environment) or its greater relative frequency. (The latter is the assumption of a weighted random environment; that is, the observations are viewed as random samples taken from a finite ensemble comprised of different numbers of each descriptor symbol.) This same prediction can also be made on the assumption of an invariant environment; that is, a constant repetition of the observed triplet symbol. It should be noted, however, that if the environment is viewed as being composed of couplets of the individual descriptor symbols, the opposite prediction would be made regardless of which pair of symbols is taken as the basis. It would appear appropriate for the predictor to examine both the relative amount of evidence that supports each hypothesis concerning the nature of the environment and the relative complexity of each hypothesis (the amount of assumed logical structure).

To summarize, to this point: The decision-maker can predict on the basis of various assumptions concerning the nature of the environment. For example, the environment may be presumed to be invariant, cyclic, inertial, or capricious; these assumptions being stated in order of decreasing severity. Taken together, these assumptions span the range from fixed to random environment.

A decision-maker can operate at a higher level of abstraction only if he has made a sequence of predictions and can recall their validity. For example, if the use of couplets is observed to consistently yield correct predictions, then it is well to base further predictions upon an interpretation of such observed data. Similarly, if most predictions of the environment prove wrong, the decision-maker may consider the possibility that his predictions were affecting the environment in some way that could successively alter its next state. On the basis of this assumption, he might choose to reverse all future predictions, assuming, of course, that the decision-maker recognizes that the environment can be in only one of two conditions; so that if his predictions were wrong the opposite predictions would necessarily be correct. (Here the decision-maker is reversing his previous hypotheses in order to make them correct and then observing the consistency among them.)

The discussion in Chapter 6 makes it evident that prediction is but a particular application of the scientific method wherein the unknown data point happens to lie in the future. In part, at least, the risk (expected loss) associated with prediction can be measured in terms of the complexity of the set of assumptions that, together with the empirical evidence, comprise the model of the environment. Only a decision-maker who has memory of previous decisions can operate at successively higher and higher levels of abstraction wherein each model is a composite of other models that have proven to be of value at a lower level of abstraction. Since all of these models were created by the decision-maker, it becomes self-evident that a wise choice of models is essential to gaining value from operating at higher levels of abstraction.

To illustrate some of these concepts, consider a simple prediction algorithm that includes a cost for the complexity of each hypothesis in terms of the inefficiency with which it represents the observed sequence of data points. For any finite sequence of observed descriptor symbols, there exists a subsequence that can completely identify an invariant or cyclic environment; that is, the observed sequence would result from the successive repetition of a subsequence. Table A.1 indicates 16 of the 32 possible observed sequences of 5-symbol length, together with a minimum subsequence which could form the basis for a cyclic environment that includes each of the observed sequences. (The remaining 16 sequences are omitted because of symmetry.) In general, a minimum-length "cyclic" subsequence can be found by comparing each successive symbol to the first symbol. If the first symbol is not encountered again, the sequence is irreducible; that is, it could only have come from a cyclic environment based upon that sequence or a longer sequence that included that sequence as a subsequence. If, however, the first symbol is found to appear a second time, the subsequence of symbols up to the repeated first symbol may be com-

Table A.1
Minimum Length Cyclic Subsequence for 5-Symbol Observations

Observed Sequence	Minimum Length Cyclic Subsequence	Length of the Model Subsequence
11111	1	1
11110	11110	5
11101	1101	4
11100	11100	5
11011	011	3
11010	11010	5
11001	1001	4
11000	11000	5
10111	0111	4
10110	110	3
10101	01	2
10100	10100	5
10011	0011	4
10010	010	3
10001	0001	4
10000	10000	5

pared with the next sequence of symbols. The comparison may be made by beginning with this repeated symbol and duplicating the first sequence as many times as is necessary to produce a sequence of length equal to that of the observed sequence. If a one-to-one correspondence exists, then a minimum subsequence has been found. If not, then one should proceed to the next symbol that is the same as the first symbol, and again compare a repetition of the subsequence up to that point with the observed sequence. This process should be repeated until a minimum-length cyclic subsequence has been found. Obviously, for any finite sequence of descriptor symbols, this procedure can take only a finite number of steps.

For example, the observed sequence 1011011 is first compared to the sequence 1010101, since 10 is the sequence of symbols that occurs before the first repetition of the initial symbol. These seven-length sequences are not identical; therefore the test sequence (10) is discarded. The next repetition of the initial symbol indicates the test sequence 101 which generates 1011011. Since this is identical with the observed sequence, the minimum cyclic subsequence has been identified to be 101.

Table A.2 indicates the number of minimum-length cyclic subsequences of each length for lengths of the observed sequence up to five descriptor symbols. It should be noted that for any sequence of length, n, greater than or equal to $2k$, the number $S_n(k)$ of minimum-length sequences of length k that generates sequences of length n is given by

$$S_n(k) = 2^k - \sum_r S_k(r)$$

r being any integer different from k which divides k.

A prediction of the environment now results from comparing the total efficiency-weight of those models that indicate each of the possible symbols to occur at the particular point in the future; these models are the minimum-length cyclic subsequences that would generate each observed subsequence of different length extending from and including the most recent symbol. The efficiency of each model is measured in terms of the relative saving in the number of symbols; that is, as the 1's complement of the ratio of the number of symbols in the model to the number of symbols in the observed sequence of symbols it represents.

Those individual efficiency-weightings descriptive of models that predict the same symbol to occur at a specific point in future time are summed. That symbol which has the greatest "vote" is then predicted to occur. This same procedure can be carried out for each point in future time. The models remain unchanged, but the fact that their individual predictions may be different results in various groupings at each point in future time. The overall prediction for each point in future time is always that symbol recommended by the greatest total efficiency; that is, that symbol which results from the simplest set of hypotheses consistent with the assumed constraints.

Table A.2

Number of Minimum Cyclic Subsequences for Lengths of Observed Sequence

Length of the Observed Sequence	Number of Minimum Cyclic Subsequences of Length 1	2	3	4	5
1	2				
2	2	2			
3	2	2	4		
4	2	2	6	6	
5	2	2	6	10	12

To clarify this prediction algorithm, consider a specific example: Let the observed sequence of descriptor symbols be 11011110011. The first column of Table A.3 indicates each of the recent subsequences of this observed sequence of symbols, proceeding from the present into the past. The second column indicates a minimum-length cyclic subsequence that, if repeated indefinitely, would generate an environment capable of including the recent subsequence as a particular observation. The next column indicates the length of each model in terms of the number of symbols in the cyclic model. The efficiency-weighting

Table A.3

Efficiency and Predictions of Minimum Length Cyclic Models

The Most Recent Subsequence	Minimum-Length Cyclic Model	Its Length	Its Efficiency*	Prediction of the Next Symbol	Prediction of Second Symbol into the Future
1	1	1	0	1	1
11	1	1	1/2	1	1
011	011	3	0	0	1
0011	0011	4	0	0	0
10011	0011	4	1/5	0	0
110011	0011	4	1/3	0	0
1110011	10011	5	2/7	1	0
01110011	01110011	8	0	0	1
101110011	01110011	8	1/9	0	1
1101110011	01110011	8	1/5	0	1

$$* 1 - \frac{\text{length of model}}{\text{length of the recent subsequence}}$$

of each model is shown in the next column in terms of the 1's complement of the ratio of the number of symbols in the representations to the number of symbols in the recent subsequence. The next column indicates the specific symbol predicted to occur next, on the basis of each cyclic model. The last column indicates the symbol predicted to occur as the second symbol into the future by each of these same models.

The overall prediction of the first symbol to occur in the future is found by summing the efficiency-weighting of those models that correspond to each "candidate symbol." Comparison of the total weights may be used to indicate not only the specific prediction of the next occurring symbol, but also the relative certainty of this prediction. The second symbol into the future may be predicted in a similar manner; that is, by summing these same weights, but in a different grouping (this being in accordance with the individual dictates for that future position). Indeed, prediction of any future symbol can be made in a similar manner, but, as might be expected, the likelihood of success for such prediction will decrease with an increase in the statistical variability of the environment. For the particular illustrative example, the next symbol into the future is predicted to be 0 on the basis of comparing 38/45 to 11/14. In a similar manner, the second symbol into the future is also to be predicted to be 0, but this by a close vote.

Note that the summing of the efficiency-weightings over the set of subsequences of different lengths extending over the observed past places a greater importance upon the more recently observed behavior of the environment. This weighting may be modified on the basis of prior information or as a result of analysis of the data. For example, a standardized variable may be computed for each subsequence that offers a measure of the probability that the given sequence might arise from the observation of a random environment.* An inverse function of the magnitude of this variable can then be used to indicate that "degree of inertia" which may be associated with each subsequence. Thus, the entire range of environments from "invariant" to "capricious" can be treated by this same prediction process.

It is of interest to examine some fundamental aspects of this prediction algorithm. A new alphabet of possible descriptor symbols is generated from the observed sequence of symbols in such a manner as to maintain maximum economy. Although it may not be immediately apparent, this new alphabet of possible descriptor symbols is weighted, in degree of utility, in terms of the "statistical" worth of each descriptor-symbol—on the basis of the available evidence (the observed sequence of symbols). That is to say, in addition to the degree of relative simplicity, something about the credibility of the various models is considered without regard for their individual differences in length. The shorter the model or the greater the number of repetitions in the sample data, the greater is the worth of the model as a descriptor of the environment and, thus, as a predictor of its future behavior. In general terms, the objective is to find an economical set of equally likely descriptors for the observed behavior of the environ-

* W. J. Dixon and F. J. Massey, Jr., *Introduction to Statistical Analysis,* McGraw-Hill, New York, 1957, p. 287.

ment. An extension of their sequence constitutes a prediction of the future behavior of the environment.

In terms of the scientific method, simpler models are preferred, but so are those models which will often demonstrate their utility. This prediction algorithm demonstrates the need for generating assumptions concerning the nature of the information source; assumptions that are consistent with, and even suggested by, the sequence of observed data. In addition, there must also be a suitable balance between the worth of modeling a greater portion of the observed data and the cost of handling such a "more complete" representation. As each new data point is observed, a new set of models can be generated. This new set may be expected to have a marked similarity to its successor, both being based upon almost the same recall. Here is an iterative means for the sequential prediction of time series that benefits from the redundancy of "inheritance." Unfortunately, as described, it also suffers from the severe restriction to the class of cyclic models.

APPENDIX B

Concerning Measures of Complexity

Obviously, to characterize the complexity of a machine in terms of the number of states which comprise the state diagram of the machine is insufficient. For instance, clearly the state-output machine described by the state diagram of Figure B.1(*a*) is more complex than the machine described in Figure B.1(*b*). What is needed is some means for describing the logical complexity of any given machine. Unfortunately, such complexity is multidimensional so that in many cases no simple ordering may exist. The problem is similar to that which occurs in the comparison of numeric matrices. It is, of course, easy to write two such matrices which are different from one another but it is difficult to introduce a meaningful ordering.

To allow the introduction of a meaningful penalty for the complexity of machines as a useful parameter in the design procedure, there must be some single measure of the overall cost of complexity; this is a composite mapping of the various aspects of complexity, each measured in terms of its cost. A first step in this direction is the identifying of these individual aspects of complexity in terms of measurable properties of machines.

An atomistic approach to this problem appears to be in order; that is, the finding of some way to represent any machine in terms of a composite of elementary machines. Referring to this decomposition, it should be possible to compare the individual complexity, the number, and the connectivity of the elementary machines, these being a common set of measures for any machine. By analogy, suppose that it is desired to measure the relative complexity of the molecules of various chemicals: The problem, in many cases, is difficult indeed,

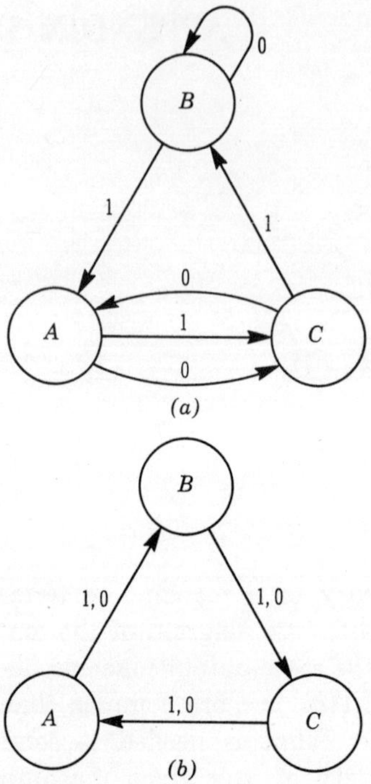

Fig. B.1. Two three-state machines of different complexity.

but, certainly, it is intuitively evident that a molecule of hydrogen peroxide, H_2O_2, is more complex than a molecule of water, H_2O, in terms of having an additional oxygen atom. Then, in a similar manner, it is easy to say that a molecule of copper sulfate, $CuSO_4$, is more complex than a molecule of magnesium sulfate, $MgSO_4$, because of the greater complexity of the copper atom. Similarly, molecules of two different organic chemicals might have the same set of atoms and still be of the same complexity in terms of the number of bonds. Lastly, there might even be two different molecules with the same atoms and the same number of bonds but different in configuration and, thus, different in complexity. In this latter regard, the problem is most difficult; and, to gain additional information, it may be necessary to return to a comparison of their relative behavioral properties by saying that the molecule is more complex which performs all the desired functions, and more, of another molecule.

As a result of work by Krohn and Rhodes on the algebraic theory of machines,* it is now possible to take a major step toward the quantitative measurement of the complexity of machines. The fundamental result of Krohn's and Rhodes's work concerns the decomposition of any given finite-state machine into constituent prime machines. More specifically, the prime decomposition of a given machine consists of that series-parallel configuration of prime machines which, with a minimum number of such machines, will behave in exactly the same manner as the given machine; that is, there exists no finite sequence of input symbols that will result in a

* K. B. Krohn and J. L. Rhodes, "Algebraic Theory of Machines I, Main Decomposition Theorem. Prime Decomposition of Finite Semigroups," *Technical Report* 63-1, Prepared under contract to Office of Naval Research, Sept. 1963.

difference in the resulting sequence of output symbols. As might be expected, a prime machine is defined as a machine that cannot be synthesied as a composite of other machines.

In order to identify the prime machines of a decomposition, it is often helpful to find the prime loops in the given machine. A loop of a machine is a sequence of distinct states, $q_1, q_2, \ldots, q_n$ such that under the imposition of a fixed input sequence t, $q_i \rightarrow q_{i+1}$ for $i = 1, \ldots, n-1$ and $q_n \rightarrow q_1$. The length of such a loop is n; that is, the number of different states in the sequence. A loop is said to be a prime loop if, and only if, the length of the loop is a prime integer. It is important to note that the prime loops of a machine *cannot,* in general, be found in terms of obvious geometrical properties of the given state diagram. Loops of length 1 lack significance and are therefore disregarded in the following discussion.

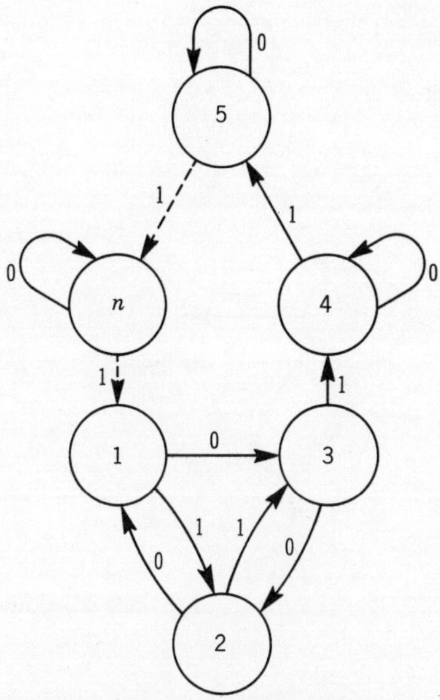

Fig. B.2. The family of machines M_n.

To illustrate, consider the family of machines, M_n, shown in Figure B.2, n being finite. For $n = 3$, the general machine reduces to the simple cyclic machine shown in Figure B.3. The prime loops of this machine are found by imposing the input symbol 0 which causes a

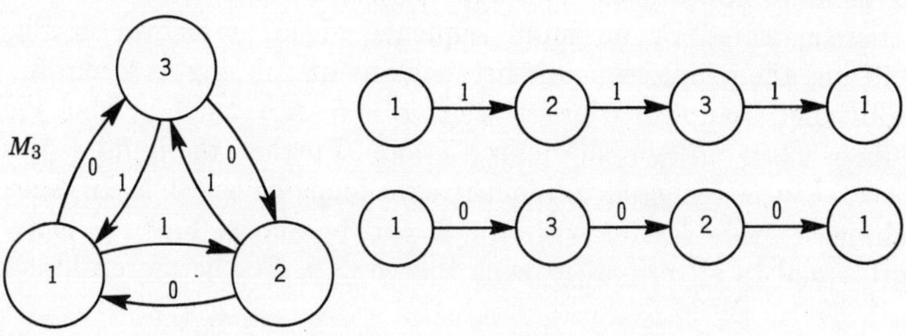

Fig. B.3. Machine M_3 and its prime loops.

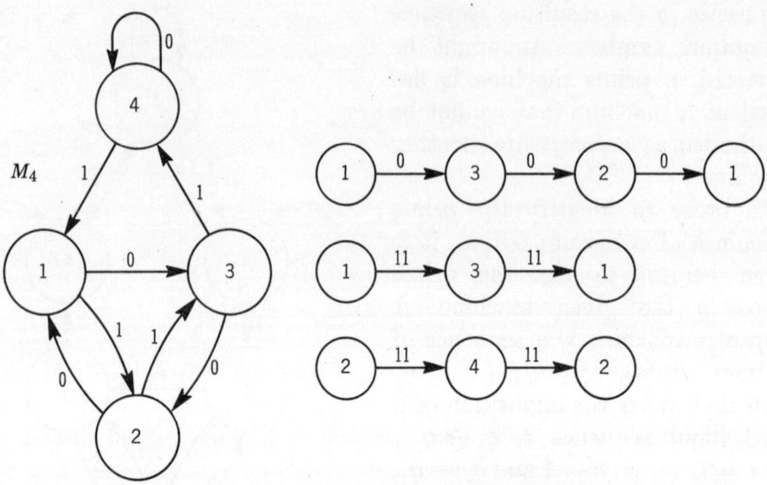

Fig. B.4. Machine M_4 and its prime loops.

cyclic displacement with a return to any initial state after three transitions. Three is a prime number; therefore, it can be said that the input of a 0 identifies a prime loop of length 3 (this being more conveniently referred to as a 3-loop). Similarly, an input of 1 identifies another 3-loop. It is easy to see that all longer sequences of input symbols only cause repetitions of these individual loops or a 1-loop.

If $n = 4$, as shown in Figure B.4, a 0 input symbol cyclically permutes the states, 1, 2, 3, as in the case of $n = 3$. Consequently, there is a 3-loop. In the same manner, a 1 input symbol cyclically permutes the four states, thus giving a loop length 4. It can be shown that if an input sequence t yields a loop of length $n_1 \cdot n_2$, then the input sequence t^{n_1} yields a loop of length n_2 (where the exponent refers to the number of repetitions of the sequence t).* As a result, a 4-loop is equivalent to two 2-loops in the prime decomposition. (These 2-loops are demonstrated by an input sequence of 11, as shown in Figure B.4. Thus, the prime loops of this machine are of lengths 2 and 3.

Additional complexity arises when $n = 5$. A 1 input symbol yields a 5-loop, a 0 input symbol yields a 3-loop. Further, the input sequence 0111101 can be imposed to identify two loops of length 2. In general, machines of this family with $n \geq 5$ can be shown to have loops of length equal to all primes, p, such that $p \leq n$. Furthermore, this fam-

* This property allows any loop to be replaced by its prime loop decomposition. For example, a 12-loop is equivalent to the existence of two 2-loops and one 3-loop.

ily belongs to the larger class of group machines (machines in which every input sequence permutes the states). These machines can have p loops for only those p less than, or equal to, the number of states n.

To fully characterize a given machine, it is sometimes useful to find its prime loops. Many of the prime loops correspond to modulo p counters in the decomposition of the machine. As long as this is true, their individual lengths form a measure of their complexity. There are also, however, prime machines which are not equivalent to counters of any modulus. The machine described earlier ($n = 5$, shown in Figure B.5) is the smallest such machine. In fact, all members of this family of machines M_n, for which $n \geq 5$, fall into this class. In general, it can be shown that if there are only prime loops of length p and q ($p \neq q$), or if there are no 2-loops, then the given machine can be sequentially decomposed into modulo-counters involving only those primes corresponding to loops of prime number length. (That is to say, if the set of the lengths of the prime loops does not contain the length 2, then the machine can be built out of counters, delays, and/or unit elements.) If these conditions are not met, prime loops are still of significant value since they either identify or leave open the possibility of more complex prime machines which can be complexity-measured in terms of their number of states and other properties.

Quite obviously, an additional aspect of complexity is the minimal number of each type of prime machines in the decomposition; and, beyond this, still greater refinement may be introduced in the form of a measure of the relative complexity associated with various configurations of the loops with respect to one another and of the required coding for transfer between the prime machines.

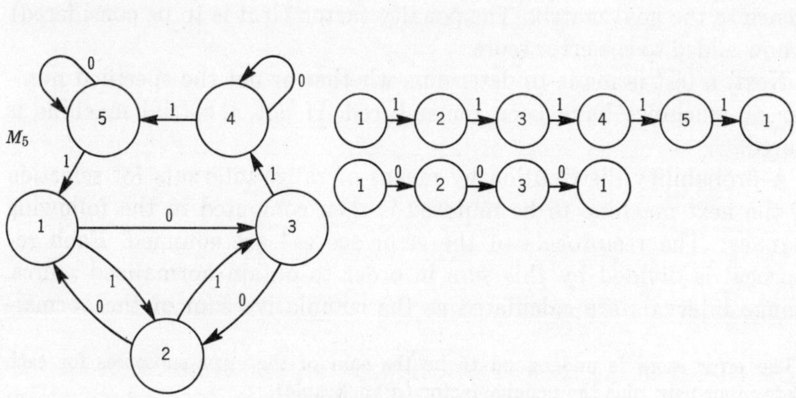

Fig. B.5. Machine M_5 and its prime loops.

APPENDIX C

Logic and Flow Charts of an Evolutionary Program

The narrative description of the logical flow of an 8-symbol evolutionary program can better be followed by referring to Figures C.1, C.2, and C.3.

To properly identify each computer run, the input data are first read in and printed out. The initial trial machine is then exercised over the recall in order that the experience gained can be utilized in deterministically setting the output for each state-input pair so that a minimum machine score over the recall is attained. This is done by determining that output which, of those required for each state-input pair, results in the smallest error subscore for each transition. Error terms are determined by reference to the goal matrix. The output for a state-input pair not exercised over the recall is set to that symbol which has so far been experienced most frequently. Next, the error score for the trial machine is computed over the recall by reference to the goal matrix. The penalty factor (if it is to be considered) is now added to the error score.

Next, a test is made to determine whether or not the specified number of machines have been remembered. If not, the trial machine is stored.

A probability distribution by means of range intervals for selection of the next machine to be mutated is then computed in the following manner: The reciprocals of the error scores* are summed. Each reciprocal is divided by this sum in order to obtain normalized scores. Range intervals are calculated as the cumulative sum of the normal-

* The error score is understood to be the sum of the error subscores for each state-input pair, plus the penalty factor (if applicable).

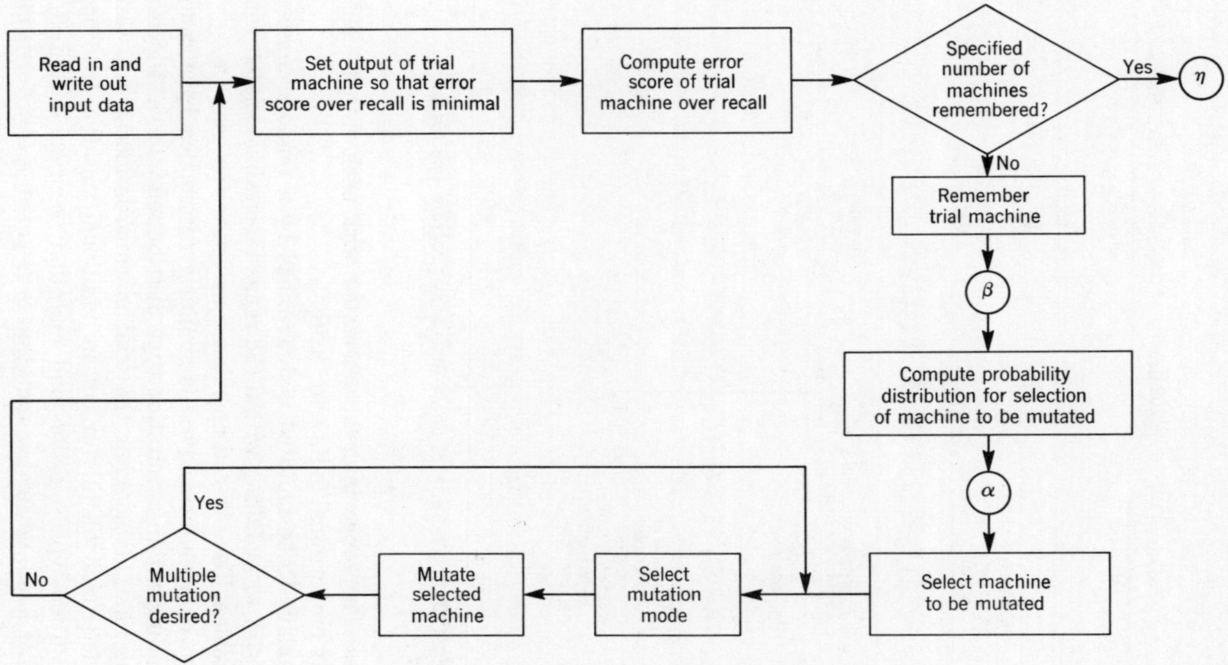

Fig. C.1. Logical flow of an evolutionary program.

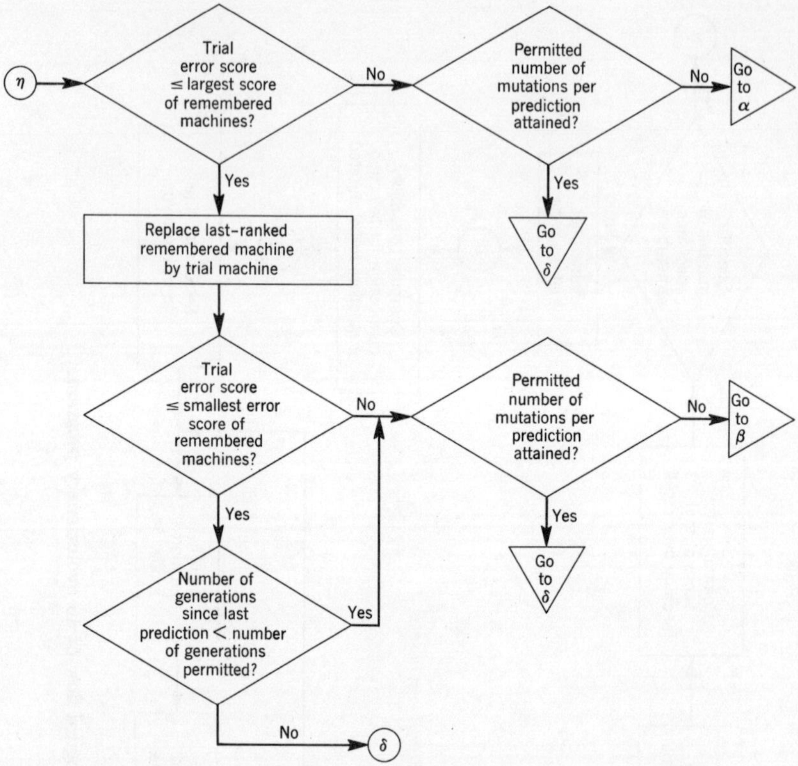

Fig. C.2. Logical flow of an evolutionary program (*continued*).

ized scores. The range interval associated with each stored machine is inversely proportional to its error score.

The machine to be mutated is determined by a random number R (where $0 \leq R < 1$) falling within the range interval associated with a stored machine. The selected machine is moved to the trial area.

Selection of a mutation type is similarly determined by a random number falling within a mutation type distribution interval which was read in as input data. After the trial machine is mutated, a test is made to determine whether or not an additional mutation is to occur. If so, mutation-type selection and mutation occur as described. If, on the other hand, no further mutation is required, a transfer is made to that portion of the program wherein the output symbol for each state-input pair is set deterministically.

Attention is returned to the test that determines whether or not the specified number of machines have been stored. Assuming the re-

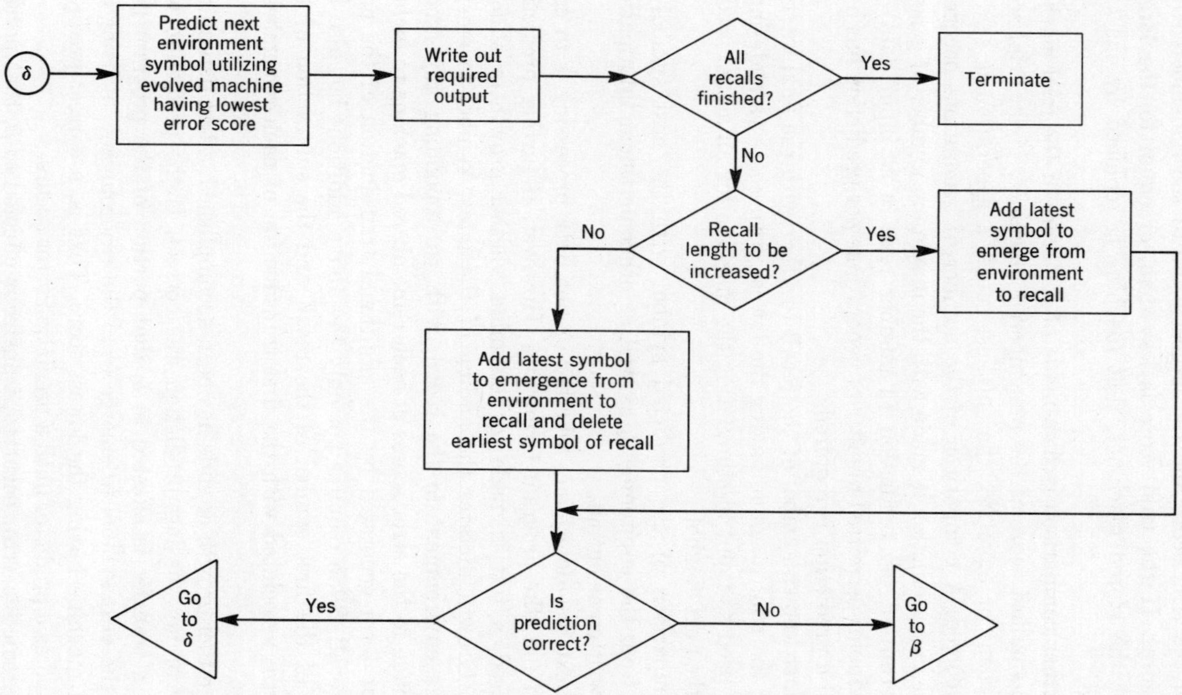

Fig. C.3. Logical flow of an evolutionary program (*continued*).

quired number of machines have been remembered, a comparison is made of the trial score and the largest score of the remembered machine scores. If the trial score is less than, or equal to, the largest score of the stored machines, that machine is replaced by the trial machine.

A further comparison is then made of the score of the trial machine and the smallest score of the remembered machines. If the trial score is less than, or equal to, the smallest score, a generation is said to have occurred. A comparison of the number of generations occurring since the last prediction is made with the maximum number of generations permitted per prediction (a number read in as input data). If the maximum permitted number of generations per prediction has been attained a prediction is required.

The last known symbol of the recall together with the final internal state of the trial machine having the lowest error score (a state-input pair) is used as input in stimulating the machine for an actual prediction of the next symbol.

A comparison of the predicted symbol with the next symbol to emerge from the environment, as well as other pertinent information, is then written as output.

The next test determines whether or not data processing is to continue. If not, the program terminates. However, if further prediction is required, a test is made to determine whether growth or slide is desired. Growth dictates the adding of the latest symbol to emerge from the environment to the recall, with corresponding adjustment being made to the error scores of each remembered machine to include the error term generated by the additional transduction of the new, longer recall. Slide requires not only the above additions but also the discard of the first symbol of the recall and the subtraction of the error term associated with the first transduction of each remembered machine.

After either growth or slide has been accomplished, a test determines whether or not the last prediction was correct. If the prediction was correct, a transfer is effected to δ, that portion of the program predicting the next symbol to emerge from the environment, utilizing the evolved machine having the lowest score. That is, a correct predictor is retained as a predictor until a prediction error occurs.

If a prediction error occurred, transfer is effected to β, that portion of the program computing the probability distribution for selection of a machine to be mutated.

Attention is directed to the test at location η. Assuming that the trial machine score is greater than the largest score of the stored machines, retention of the trial machine is not required, and is therefore discarded. A test is then made to determine whether the maximum permitted number of mutations per prediction has been attained. If not, a transfer is effected to α, that portion of the program wherein a machine is selected for mutation. On the other hand, if the maximum permitted mutations per prediction have been attained, transfer is effected to δ.

Note should be made again of the comparison of the trial error score with that of the smallest score of the remembered machines. Assuming the trial error score is the larger, a next test determines whether or not a maximum number of permitted mutations has been attained. If so, a transfer is effected to δ; otherwise, a transfer is effected to β.

Consider again the test for the comparison of the number of generations since the last prediction with the maximum number of generations permitted per prediction. Assuming the maximum has not been attained, a transfer is effected to a test of the number of mutations per prediction attained, and the logic proceeds as before.

APPENDIX D

Finite-State Machines as Primitive Recursive Functions

As previously noted, for the output of a finite-state machine to perfectly predict an environment, the environment must be cyclic; that is, it must consist of a finite sequence $a_0a_1 \cdots a_{k-1}$, repeated indefinitely. If we represented this cyclic sequence as a function, f, on the integers, we would have

$$\begin{aligned} f(n) &= a_n \quad \text{for } 0 \leq n < k \\ &= a_r \quad \text{for } n \geq k, \text{ where } r \text{ is the remainder when } n \text{ is divided by } k \end{aligned}$$

Hence, to show that the function defined by a finite-state machine is primitive recursive, it suffices to indicate that such a cyclic function, f, is primitive recursive.

As shown by Davis,[7] the functions $x + y$, xy, and $x \dot{-} y$ are primitive recursive. Let a predicate $P(y, x)$ be termed primitive recursive if the characteristic function of the set of values for which the predicate is true is primitive recursive. Let $P(y, x)$ be a predicate and $\underset{y=0}{\overset{z}{M}} P(y, x)$ be the function whose value for a given x and z is the least $y \leq z$, if it exists, for which $P(y, x)$ is true. If no such y exists, then the value of the function is zero. Then the function $\underset{y=0}{\overset{z}{M}} P(y, x)$ is primitive recursive if the predicate $P(y, x)$ is primitive recursive[8]:

A proof that the cyclic function, f, is primitive recursive proceeds by a list of functions and predicates shown to be primitive recursive:

$g(x) = a$ for all x is primitive recursive.

[7] M. Davis, *Computability and Unsolvability,* McGraw-Hill, New York, 1958.
[8] See Davis, *ibid.*

Proof: $g(x) = S^a(N(x))$ where by S^a is meant the a-fold composition of the function S; that is, $S^1(x) = S(x)$, $S^2(x) = S(S^1(x))$, . . . , $S^a(x) = S(S^{a-1}(x))$. It should be recalled that $S(x) = x + 1$ and $N(x) = 0$ for all x.

The predicate $ky + k > n$ is primitive recursive, where k is a fixed integer.

Proof: The characteristic function for this predicate is

$$1 \dot{-} (1 \dot{-} ((ky + k) \dot{-} n))$$

The function $[n/k]$ = largest integer $\leq n/k$ is primitive recursive.

Proof:

$$[n/k] = \underset{y=0}{\overset{n}{M}} \; (ky + k > n)$$

The function $R(n, k)$ = remainder when n is divided by k is primitive recursive.

Proof:

$$R(n, k) = n \dot{-} k[n/k]$$

The function $|x - y|$ is primitive recursive.

Proof:

$$|x - y| = (x \dot{-} y) + (y \dot{-} x)$$

The function $C_n(x)$, that is, the characteristic function of the set whose only member is the integer n, is primitive recursive.

Proof:

$$C_n(x) = 1 \dot{-} |n - x|$$

If $a_0, a_1 \cdots, a_{k-1}$ is any set of k integers, then the cyclic function $f(n) = a_r$ where r is the remainder when n is divided by k is primitive recursive.

Proof: The k functions f_r, $0 \leq r < k$, defined by $f_r(x) = a_r$ for all x are primitive recursive. $C_r(R(n, k))$, being the composition of primitive recursive functions, is primitive recursive for each r, $0 \leq r < k$. Since the sum or product of primitive recursive functions is primitive recursive, the function

$$F(n) = \sum_{r=0}^{k-1} f_r(n) C_r(R(n, k))$$

is primitive recursive. Now $C_r(R(n, k))$ is 0 if $r \neq R(n, k)$, and is 1 if $r = R(n, k)$; thus, $F(n) = f_r(n) = a_r$ where r is the remainder when n is divided by k. However, this is exactly $f(n)$, so that $F = f$ and f is primitive recursive.

The Turing machine given by Table D.1 is one which, when started in state q_1, completely describes the sequence 101100111000 · · · in that if it is started in state q_1 with any input in the scanned position and blanks in all other positions of the tape, it will print out this sequence. Basically, this Turing machine is a set of quintuples of the form $q_iS_jS_kRq_m$ or the form $q_iS_jS_kLq_m$. These are interpreted as follows: If the machine is in internal state q_i with observed symbol S_j, then the symbol S_k is printed on the tape (if S_k = blank symbol, that position is erased), the tape is moved right (or left) one position and the machine enters internal state q_m.

Table D.1

	0	1	*	Blank
q_1	$1Rq_2$	$1Rq_2$	$1Rq_2$	$1Rq_2$
q_2				$0Rq_3$
q_3				$1Rq_4$
q_4				$1Rq_5$
q_5	$0Lq_5$	$1Lq_6$		$*Lq_5$
q_6	$0Lq_7$	$1Lq_6$	$0Rq_8$	
q_7	$0Lq_7$	$1Rq_8$	$0Rq_8$	
q_8	$*Rq_9$	$1Rq_{10}$		
q_9	$0Rq_9$	$1Rq_9$	$0Rq_5$	
q_{10}	$0Rq_{10}$	$1Rq_{10}$	$0Rq_{10}$	$*Lq_{11}$
q_{11}	$0Lq_{12}$	$1Lq_{11}$		
q_{12}	$0Lq_{12}$	$1Lq_{13}$	$1Rq_{14}$	
q_{13}	$0Rq_{14}$	$1Lq_{13}$	$1Rq_{14}$	
q_{14}	$0Rq_{16}$	$*Rq_{15}$		
q_{15}	$0Rq_{15}$	$1Rq_{15}$	$1Rq_{10}$	
q_{16}	$0Rq_{16}$	$1Rq_{16}$	$1Rq_{16}$	$*Lq_5$

In comparison with the quadruple representation, $q_iS_jS_kRq_m$ would be equivalent to a pair of quadruples, $q_iS_jS_kq_r$ and $q_rS_kRq_m$. A similar equivalence holds for $q_iS_jS_kLq_m$. For a complete representation of Turing machines, two other types of quintuples would be required: $q_iS_jS_kNq_m$, where N means no tape movement, and $q_iS_jS_kq_rq_m$, where transfer is to state q_r if S_j is in a given set A, and to state q_m if S_j is not in the set A.

In Table D.1 the current state q_i is given in the left-hand column, and the scanned symbol is given in the top row. Entries in the corresponding block of the table are the last three members of the quintuple.

The operation of the machine is indicated by listing the successive quintuples encountered, and describing the corresponding activities of the machine and tape. (A blank is indicated by b; $*$ is the symbol used as a marker, and x indicates that any symbol may be inserted.)

1. q_1x1Rq_2 Prints out 1 in position 1; moves to the right and enters state q_2.
2. q_2b0Rq_3 Prints out 0 in position 2; moves to the right and enters state q_3.
3. q_3b1Rq_4 Prints out 1 in position 3; moves to the right and enters state q_4.
4. q_4b1Rq_5 Prints out 1 in position 4; moves to the right and enters state q_5.
5. q_5b*Lq_5 Prints out $*$ in position 5; moves to the left and remains in state q_5.
6. q_611Lq_6 Leaves 1 in position 4; moves to the left and enters state q_6.
7. q_611Lq_6 Leaves 1 in position 3; moves to the left and remains in state q_5.
8. q_600Lq_7 Leaves 0 in position 2; moves to the left and enters state q_7.
9. q_711Rq_8 Leaves 1 in position 1; moves to the right and enters state q_8.
10. q_80*Rq_9 Replaces 0 by $*$ in position 2; moves to the right and enters state q_9.
11. q_911Rq_9 Leaves 1 in position 3; moves to the right and remains in state q_9.
12. q_911Rq_9 Leaves 1 in position 4; moves to the right and remains in state q_9.
13. q_9*0Rq_5 Replaces $*$ by 0 in position 5; moves to the right and enters state q_5.
14. q_5b*Lq_5 Prints out $*$ in position 6; moves to the left and remains in state q_5.
15. q_500Lq_5 Leaves 0 in position 5; moves to the left and remains in state q_5.

Steps 5, 6, and 7 are repeated, but step 8, since there is now an asterisk rather than a 0 in position 2, is replaced by

16. q_6*0Rq_8 Replaces $*$ by 0 in position 2; moves to the right and enters state q_8.
17. q_811Rq_{10} Leaves 1 in position 3; moves to the right and enters state q_{10}.

While in state q_{10}, the machine leaves all zeros and ones unchanged, moves to the right and remains in state q_{10}. It therefore moves right to position 6 leaving the tape unchanged. Since position 6 has an asterisk in it, the next step is:

18. $q_{10}{*}0Rq_{10}$ Replaces $*$ by 0 in position 6; moves to the right and remains in state q_{10}.

The print out is now 101100 with the machine in state q_{10} and scanning position 7 which is blank. The net result is that the machine by using $*$ as a marker has printed after the last sequence of 1's a sequence of 0's of length 1 greater than the length of the sequence of 0's in front of the 1's. The next quintuple is then $q_{10}b{*}Lq_{11}$. State q_{11} through q_{16} are similar to states q_5 through q_{10} except that the symbol $*$ is now associated with the symbol 1 rather than 0. The result is that the machine by using the asterisk as a marker prints out after the last sequence of 0's a sequence of 1's of length 1 greater than the preceding sequence of 1's.

The machine is then in state q_{16} with a blank in the scanned position so that the relevant quintuple is $q_{16}b{*}Lq_5$. After this step, the machine is back in step 5 so that the computation is repeated with alternating sequences of 0's and 1's, each increased in length by 1, printed after the current sequence.

APPENDIX E

Alternative Representations for Finite-State Machines

Choice of the class of functions within which to express the organism is only part of the problem. Even within a particular class the choice of expression for the organism may have a significant effect upon the nature of the kinds of mutation which are possible. To see that this is true, consider some of the more "popular" forms of expression for finite-state machines.

Probably the simplest of these is the *state-diagram.* Here each of the possible states of the machine is represented as a node. Each of the possible transitions which can occur upon the receipt of an input symbol is indicated in the form of a directed branch connecting the appropriate nodes. A short arrow indicates the initial state. To illustrate, consider the simple binary triplet indicator which outputs a 1 upon the receipt of every third input 1. This machine is represented by the state-diagram shown in Figure E.1. Here the symbol before each virgule indicates the input symbol causing that particular transition, the produced output symbol being indicated after the virgule. Note that there are five simple and independent ways in which a state-diagram can be randomly modified: replace an arbitrarily chosen output symbol by another symbol, change the initial state, change the next state of an arbitrarily chosen state transition, add a randomly connected additional state, and delete a randomly chosen state (randomly distributing the input transitions to that state among the remaining states). The rules of mutation can include any or all of these modes of modification (or multiples thereof) since the resultant state-diagram will still represent some finite-state machine.

For the sake of completeness is it worthwhile to note the possibility

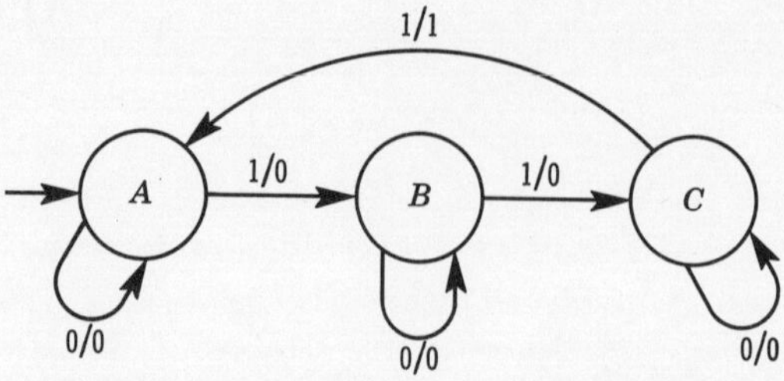

Fig. E.1. State-diagram of a triplet counter.

of expressing any finite-state machine in terms of its state-output diagrams. The symbol associated with an arrow results in the transition from the state at the tail of the arrow to the state at the head of the arrow, whereas the symbol to the right of the colon within each state refers to the output symbol effected upon entering that state. In such an expression each state corresponds only to a single output symbol. Thus it is usual to find that the state-output representation for a particular machine has a greater number of states than those contained in the state diagram. Figure E.2 offers an equivalent state-output expression for the above referenced triplet indicator. Note that the initial state can be either *A* or *A′*. The modes of mutation for such a representation are different from those listed above, in that it is no longer possible to change a randomly chosen output symbol.

Finite-state machines may also be expressed in terms of their equiv-

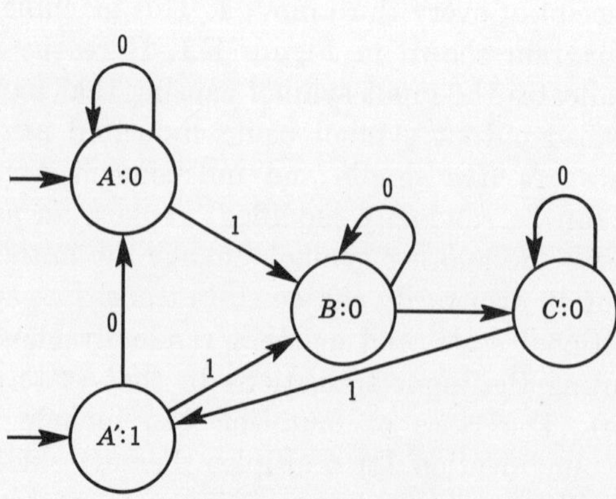

Fig. E.2. State output diagram of a triplet counter.

alent semigroup representation. The semigroup may be viewed as a description of the transduction on the machine in terms of the transformation it applies to that set of input sequences which result in distinguishably different operations of the machine. This includes not only the outputs but also the internal state changes. To find the semigroup of a machine, consider each of the possible input sequences of increasing length in terms of the resulting final state and associated output symbol (this applies to each of the states of the machine). Table E.1 offers just such an expression for the triplet indicator. Comparison

Table E.1
Characteristic Response of the Triplet Indicator to Various Input Sequences

Input Sequence	*A*	*B*	*C*
0	*A*0	*B*0	*C*0
1	*B*0	*C*0	*A*1
00	*A*0	*B*0	*C*0
01	*B*0	*C*0	*A*1
10	*B*0	*C*0	*A*0
11	*C*0	*A*1	*B*0
000	*A*0	*B*0	*C*0
001	*B*0	*C*0	*A*1
010	*B*0	*C*0	*A*0
011	*C*0	*A*1	*B*0
100	*B*0	*C*0	*A*0
101	*C*0	*A*1	*B*0
110	*C*0	*A*0	*B*0
111	*C*1	*B*0	*C*0

of the characteristic behavior of the machine over various sequences of input symbols allows identification of the semigroup equivalence sequences. The table need be continued only until all possible response behavior of the machine has been found. Note that for the triplet indicator all response patterns to input sequences of four symbols in length are already contained in the previous portion of the table. As a matter of convention, the elements of the semigroup are taken to be those input sequences of minimum length which produce distinguishably different response of the machine.

The semigroup itself can then be constructed in the form of a table which describes the operation of the machine in terms of its transformation of all possible input sequences, this in terms of its operation on the equivalent succession of the elements. Table E.2 represents the semigroup for the triplet indicator. Response of this machine to the sequence 101110 can be found by noting that this sequence is comprised of the elements 101 and 110. 101 is equivalent to 11 so (101) (110) = (11) (110) which, by Table E.2, is 10. The response

Table E.2
Semigroup for the Triplet Indicator

	0	1	10	11	110	111
0	0	1	10	11	110	111
1	10	11	110	111	0	1
10	10	11	110	111	0	1
11	110	111	0	1	10	11
110	110	111	0	1	10	11
111	0	1	10	11	110	111

is then given by Table E.1 in the row and column corresponding to 10, that is, 0.

Modification of a semigroup raises some fundamental problems, in that a change of the operation for any combination of elements may result in an operation which cannot be described by a semigroup in that the required associativity may have been lost. Since random modification of the semigroup representation of a finite-state machine does not, in general, produce another finite-state machine, this mode of expression is unsuitable as a basis for mutation.

Each finite-state machine may be expressed in the form of a unique number. Such a numbering system was first constructed by Gödel in his classic dissertation of 1931. To construct a Gödel number for any given machine, it is only necessary to associate a unique set of numerical descriptors to each of the input symbols and each of the states of the machine. For example, a Gödel number for the triplet indicator can be found by adopting the numerical equivalence wherein the input symbol 0 corresponds to 3, the input symbol 1 corresponds to 5, and the states of the machine are represented by $2i + 5$, wherein i is the numerical order of the state. Thus, state A corresponds to 7, state B corresponds to 9, and state C corresponds to 11, Table E.3 indicates the composite structure by which such a Gödel number can be written for any given finite-state machine. As shown, a Gödel number for the triplet indicator is

$$2^{2^7 3^3 5^7 7^3}\ 3^{2^7 3^5 5^9 7^3}\ 5^{2^9 3^3 5^9 7^3}\ 7^{2^9 3^5 5^{11} 7^3}\ 9^{2^{11} 3^3 5^{11} 7^3}\ 11^{2^{11} 3^5 5^7 7^5}.$$

The magnitude of such a number for even a relatively simple machine suggests practical difficulties which might be encountered in adopting such a mode of expression for the evolving organism.

It is, of course, possible to mutate the Gödel expression for any finite-state machine, however, it would be necessary to ensure that this mutation would produce another Gödel number. It would appear that such a change has little likelihood of producing an offspring which carries forward any meaningful measure of inheritance.

Still another representation for finite-state machines can be found in terms of their equivalent prime decomposition. According to the recent work of Krohn and Rhodes any finite-state machine can be

Table E.3
The Gödel Numbering of the Triplet Indicator

Present State	Input Symbol	Next State	Output Symbol	
A	0	A	0	$\sim 2^7\ 3^3\ 5^7\ 7^3$
A	1	B	0	$\sim 2^7\ 3^5\ 5^9\ 7^3$
B	0	B	0	$\sim 2^9\ 3^3\ 5^9\ 7^3$
B	1	C	0	$\sim 2^9\ 3^5\ 5^{11}\ 7^3$
C	0	C	0	$\sim 2^{11}\ 3^3\ 5^{11}\ 7^3$
C	1	A	1	$\sim 2^{11}\ 3^5\ 5^7\ 7^5$

written as a series-parallel combination of elementary machines. This opens the possibility of modes of mutation in terms of alternative composition and arrangement of the prime elements which comprise the original machine and additional elements. The triplet indicator is not a prime machine, but can be decomposed into the parallel composition of two prime machines starting with the semigroup $S = \{0, 1, 10, 11, 110, 111\}$ with the semigroup operation given by Table E.2. In terms of the Krohn-Rhodes decomposition, $S = R_2 \times Z_3$ where $R_2 = \{0, 1\}$ with operation $a \circ c = c$, and Z_3 is the additive group of integers modulo 3, that is, $Z_3 = \{0, 1, 2\}$ with operation $b \circ d = r$ where r is the remainder when $b + d$ is divided by 3. For element (a, b) and (c, d) in $R_2 \times Z_3$, then $(a, b)\ (c, d) = (c, r)$.

The operation table for $R_2 \times Z_3$ is shown in Table E.4. A comparison shows that this table is isomorphic with Table E.2 for S under the correspondence

$$(0,0) \sim 111,\ (1,0) \sim 0,\ (0,1) \sim 1, (1,1) \sim 10,\ (0,2) \sim 11,\ (1,2) \sim 110.$$

Here again, it is possible to define various modes of mutation, but it would appear that the problem of storing the required elementary machines and appropriate codings, plus the complexity of the mutations, would make this less desirable than using the state-diagram representation.

Considerable literature is devoted to representing finite-state machines in terms of their equivalent regular expressions. Regular expressions can be defined in a recursive manner: 1. any finite set of sequences

Table E.4
Operation Table for the $P_2 \times Z_3$ Semigroup

	(0,0)	(1,0)	(0,1)	(1,1)	(0,2)	(1,2)
(0,0)	(0,0)	(1,0)	(0,1)	(1,1)	(0,2)	(1,2)
(1,0)	(0,0)	(1,0)	(0,1)	(1,1)	(0,2)	(1,2)
(0,1)	(0,1)	(1,1)	(0,2)	(1,2)	(0,0)	(1,0)
(1,1)	(0,1)	(1,1)	(0,2)	(1,2)	(0,0)	(1,0)
(0,2)	(0,2)	(1,2)	(0,0)	(1,0)	(0,1)	(1,1)
(1,2)	(0,2)	(1,2)	(0,0)	(1,0)	(0,1)	(1,1)

is a regular expression. 2. If P and Q are regular expressions, then so are $P \cup Q$, $P \cdot Q$, and P^*, where $P \cdot Q$ is the set consisting of all possible concatenations of a sequence from P and a sequence from Q, in that order; and where P^* is the set consisting of the concatenation of any finite number of sequences from P together with the null sequence, ϕ. A regular expression for the triplet counter is $(0^* \cdot 1 \cdot 0^* \cdot 1 \cdot 0^* \cdot 1)^*$.

But this list of alternative expressions for finite-state machines is far from complete. For example, a logical flow diagram for a triplet indicator can be drawn to include logical operators. The schematic diagram of the electronic components is another representation of the same machine. Still another representation is comprised of the components themselves. Each of these modes of representation may be randomly modified to produce offspring.

In order to select among these modes of expression it is well to note that they fall naturally into certain categories. First, there are those which describe the organism uniquely but offer almost no characterization of its operational properties. Gödel numbering systems fall within this category. Then, there are those modes of expression which define the organism in terms of its response to certain sequences of input symbols. Semigroup representations and regular expressions fall within this category. Lastly, there are those modes of expression which define the organism in terms of its internal properties. State-diagrams, state-output diagrams, logical flow diagrams, and circuit schematics fall within this category. Of these the state-diagram offers a most compact expression of only the relevant information.

The choice among these three categories must rest upon the nature of the evolutionary process and, in particular, on the desired nature of the inheritance. At least in biological evolution it would appear that inheritance exists in the form of those latent stimulus-response properties which are transmitted intact from parent to offspring. Certain gross modes of behavior are structured within instinct so that the organism, once it adopts a mode such as fight, flight, search for food, sleep, and so forth, will respond to each stimulus in a characteristic manner. In essence, the overall inheritance may be represented by a state-diagram wherein each mode of behavior corresponds to a state defined by the transitions from that state and the related output symbols to each of the possible stimuli. With the analogy of the evolutionary program to the process of biological evolution in mind, it would appear most reasonable to adopt the state-diagram representation for the evolving organism.

APPENDIX F

Logical Processes of the Scientific Method

The scientific method requires the logical processes of induction and deduction. The distinction between these can be made clear. Deductive manipulation can be carried out with complete certainty. In the classical illustration, the premise is "Socrates is a man." The deterministic rule is "All men are mortal." Therefore, "Socrates is mortal."

Often, it occurs that a particular causal factor is to be estimated on the basis of some given outcome effect and a governing rule. The classical idiom, "Socrates is mortal, all men are mortal, therefore Socrates is a man," may be true; but Socrates might also be a horse. Even though uncertainty prevails, a decision-maker may have no better course of action than to use the given rule in such an inverse manner. There is essential uncertainty, and yet this process is relied upon in everyday experience. Possibly the greatest danger in doing this lies in not realizing the degree of uncertainty that is inherent in the conclusion.[1]

Induction is a process of generalization resulting in an increase in the amount of information over that contained in the sensed data. To illustrate: "Socrates is a man, Socrates is mortal," therefore, the conclusion, "All men are mortal." It is evident that this need not be the case solely on the basis of these two observations. The result of induction can be any of an infinite set of rules that might fit the

[1] Evidence is presented. The jury is presumed to be sufficiently mature and experienced in patterns of human behavior. Based upon these "rules" and the provided evidence, these "good men and true" must determine the most likely causal factor and declare innocence or guilt accordingly. Circumstantial evidence has certainly been known to lead to false conclusions.

given data. The investigator must select a particular model from the set of possible models and carry out the remaining steps of the scientific method in order to confirm or deny its validity.

To be useful, a model must include a greater number of data points than the available evidence upon which it is based. Unfortunately, the inferred estimate of the unknown must result from just this additional information (otherwise, it would not be unknown in the first place). The initial construction of a model of an unknown environment requires the introduction of "noise;" that is, the addition of information having no intentional dependency upon the available empirical evidence.[2] However, as the induction-inference-evaluation process is repetitively applied, the required additional information can, in part, be drawn from the collection of models that have already proven to be successful estimators of the environment. In fact, the entire sequence of models that have already been verified can contribute to the generation of each new model, even though their individual acceptance was based upon the verification of but a single estimated data point. It is in this manner that each suitable model contributes multiply to the success of the exploration.

The key to successful induction, then, lies in the addition of suitable information, this being constrained to the choice of an appropriate set of assumptions concerning the environment. Such assumptions, derived from previously successful models, are usually stated in terms of descriptor symbols at the next higher level of abstraction. To illustrate, if, of the variety of models that have been used for estimating the behavior of an environment, only exponential models have been verified, it becomes reasonable to assume that the next model should also be of this form. Additional assumptions might be derived from observable trends in the sequence of previously verified models. In this manner, at each point in time a new model is constructed that represents a composite of the known data points and a set of assumptions that indicate some underlying consistency among those models that have already demonstrated the validity of their inference.

But the problem of selecting a set of suitable assumptions may not be simple. The assumptions that contributed to the generation of the previously verified models may be somewhat inconsistent. The choice among these assumptions is then complicated by the relative number of times each assumption was part of a model that proved to be "true."

[2] The finiteness of the decision-making entity precludes perfect randomness of the noise. Further, it is recognized that the structure and organization of the entity may influence both what it observes and its generation of random data.

In real-world exploration, the degree of truth of an estimate is modulated by the observed error. This introduces another bias in the estimated worth of each assumption. If the real world is inertial, would it not be better to place greater worth on those assumptions more recently verified?

Further, there is the question of the proper number of assumptions. A relatively well-behaved environment will offer consistent verification of models, based on a certain set of assumptions. Under such conditions, the investigator may find it desirable to introduce additional randomness in terms of arbitrary assumptions in the hope of gaining a deeper insight into the nature of the environment; the greater risk being justified by the anticipated greater payoff. On the other hand, the exploration of certain environments may result in the verification of widely dissimilar models. The investigator may then find a need for reducing the level of uncertainty. He may choose to explore the validity of individual assumptions rather than the conglomerate composite "suggested" by the set of previously verified models. The investigator may prefer to examine the environment by using simple models until the consistency of verification justifies his increasing their complexity.

Generally, simpler models that contain the available empirical evidence (or other acceptable data) are of greater worth. This widely accepted dictum is usually justified in terms of allowing a greater number of iterations of induction-inference-evaluation within the available time and energy, thereby increasing the likelihood of discovering models that are verified. The result is likely to be a greater penetration into the unknown. This concept is certainly not new. The maxim of parsimony, "Occam's Razor,"[3] has withstood the test of time.

In summary, inductive generalization is carried out on the basis of some available evidence that is represented in the form of symbols chosen from a finite alphabet of possible descriptors. These symbols are then embedded within a desired degree of uncertainty, in such a way as, hopefully, to preserve the evidence and, at the same time to display its complications in a readily interpretable manner. In other words, induction requires the addition of new information, but in such a manner as to minimally disturb the sensed data.

[3] This principle was propounded by William of Occam in the 14th century.

Bibliography

Anfinsen, C. B., *The Molecular Basis of Evolution,* John Wiley and Sons, New York, 1959.

Arrow, K. J., "Alternative Approaches to the Theory of Choice in Risk-Taking Situations," *Econometrica,* Vol. 19 (1951), pp. 404–437.

Ashby, W. R., *Design for a Brain,* John Wiley and Sons, New York, second ed., 1960.

Atkinson, R. C. (editor), *Studies on Mathematical Psychology,* Stanford University Press, Stanford, Calif., 1964.

Barker, S. F., *Induction and Hypothesis,* Cornell University Press, Ithaca, New York, 1957.

Bazilevskii, Yu. YA. (editor), *The Theory of Mathematical Machines,* Pergamon Press, The Macmillan Co., New York, 1963.

Bellman, R., *Adaptive Control Processes: A Guided Tour,* Princeton University Press, Princeton, N. J., 1961.

Bellman, R., *Dynamic Programming,* Princeton University Press, Princeton, N. J., 1957.

Bionics Symposium—Living Prototypes, The Key to New Technology, USAF WADD TR 60-600 (September 1960).

Birkhoff, G. D., *Aesthetic Measure,* Harvard University Press, Cambridge, Mass., 1933.

Blackwell, D., and M. A. Girshick, *Theory of Games and Statistical Decisions,* John Wiley and Sons, New York, 1954.

Blake, D. V., and A. M. Uttley (editors), "Mechanisation of Thought Processes," *Proceedings of a Symposium held at the National Physical Laboratory on 24–27 November 1958,* Vols. I and II, published by Her Majesty's Stationery Office, London, 1959.

Blum H., *Time's Arrow and Evolution,* Princeton University Press, Princeton, N. J., 1951.

Bruner, J. S., J. J. Goodnow, and G. A. Austin, *A Study of Thinking,* John Wiley and Sons, New York, 1956.

Burks, A. W., and H. Wang, "Logic of Automata, Part I & II," *Journal of Association for Computing Machinery,* Vol. 4 No. 2, and Vol. 4 No. 3 (April and July 1957), 193–218 and 279–297.

Bush, R. R., and F. Mosteller, *Stochastic Models for Learning,* John Wiley and Sons, New York, 1955.

Carnap, R., *Meaning and Necessity,* University of Chicago Press, Chicago, 1947.

Carnap, R., "The Nature and Application of Inductive Logic," taken from *Logical Foundation of Probability,* University of Chicago Press, Chicago, 1950.

Carnap, R., "Testability and Meaning," reprinted from *Philosophy of Science,* Vol. II, 1936, and Vol. IV, 1937. Whitlock's, New Haven, 1954.

Churchman, C. W., *Prediction and Optimal Decision; Philosophical Issues of a Science of Values,* Prentice-Hall, Englewood Cliffs, N. J., 1961.

Cox, R. T., *The Algebra of Probable Inference,* Johns Hopkins Press, Baltimore, Md., 1961.

Davidson, D., and P. Suppes, *Decision-Making, and Experimental Approach,* Stanford University Press, Stanford, Calif., 1957.

Davis, M., *Computability and Unsolvability,* McGraw-Hill, New York, 1958.

de Finetti, "Foresight: Its Logical Laws, Its Subjective Sources," pp. 93–158 in *Studies in Subjective Probability,* edited by H. E. Kyburg, Jr., and H. E. Smokler, John Wiley and Sons, New York, 1964.

Dekker, J. C. E. (editor), "Recursive Function Theory," *Proceedings of Symposia on Pure Mathematics,* Vol. V, American Mathematical Society, Providence, R. I., 1962.

Dobzhansky, T., *Evolution, Genetics, and Man,* John Wiley and Sons, New York, 1955.

Edwards, W., "Behavioral Decision Theory," *Annual Review of Psychology,* Vol. 12 (1961), pp. 473–498.

Edwards, W., "The Theory of Decision-Making," *Psychological Bulletin,* Vol. 51 (1954), pp. 380–417.

Fair, C. M., *The Physical Foundations of the Psyche,* Wesleyan University Press, Middletown, Conn., 1963.

Feigenbaum, E. A., and J. Feldman (editors), *Computers and Thought,* McGraw-Hill, New York, 1963.

Fogel, L. J., A. J. Owens, and M. J. Walsh, "An Evolutionary Prediction Technique," presented before the International Conference on Microwaves, Circuit Theory, and Information Theory, IEEE, Tokyo, Japan, Sept. 7–11, 1964. Summary published in Part 3 of the *Proceedings,* pp. 173–174.

Fogel, L. J., A. J. Owens, and M. J. Walsh, "Artificial Intelligence through a Simulation of Evolution," pp. 131–156 in *Biophysics and Cybernetic Systems,* edited by A. Callahan, M. Maxfield, and L. J. Fogel, Spartan Books, Washington, D. C., 1965.

Fogel, L. J., *Biotechnology: Concepts and Applications,* Prentice-Hall, Englewood Cliffs, N. J., 1963.

Fogel, L. J., A. J. Owens, and M. J. Walsh, "Intelligent Decision-Making Through a Simulation of Evolution," *IEEE, Transactions of the Professional Technical Group on Human Factors in Electronics,* Vol. 6, No. 1, pp. 13–23.

Fogel, L. J., "Levels of Intelligence in Decision Making," *Human Decisions in Complex Systems,* Annals of the New York Academy of Sciences, Vol. 89, Art. 5, pp. 732–751.

Fogel, L. J., A. J. Owens, and M. J. Walsh, "On the Evolution of Artificial Intelligence, *Proceedings of the Fifth National Symposium on Human Factors in Electronics,* IEEE, San Diego, May 5–6, 1964, pp. 63–76.

Foulkes, J. D., "A Class of Machines Which Determine the Statistical Structure of a Sequence of Characters," IRE WESCON *Convention Record,* Part 4 (1959).

Fox, J. (editor), *Mathematical Theory of Automata,* Polytechnic Press of the Polytechnic Institute of Brooklyn, New York, 1963.

Fraenkel-Conrat, H., *Design and Function at the Threshold of Life: The Viruses,* Academic Press, New York, 1962.

Friedberg, R. M., "A Learning Machine: Parts I and II," *IBM Journal of Re-*

search and Development, Vol. 2 No. 1 (January 1958), pp. 2–13, and Vol. 3 No. 7 (July 1959), pp. 282–287.

Furstenberg, H., "Stationary Processes and Prediction Theory," *Annals of Mathematics Studies,* Number 44, Princeton University Press, Princeton, N. J., 1960.

Garvin, P. (editor), *Natural Language and the Computer,* McGraw-Hill, New York, 1963.

Gill, A., *Introduction to the Theory of Finite State Machines,* McGraw-Hill Book Co., New York, 1962.

Ginsburg, S., *An Introduction to Mathematical Machine Theory,* Addison-Wesley Publishing Co., Reading, Mass., 1962.

Glushov, V. M., *Introduction to Cybernetics,* Academy of Sciences of the Ukraine, USSR (in Russian), 1964.

Hadamard, J., *The Psychology of Invention in the Mathematical Field,* first edition published by Princeton University Press, 1945, reprinted, Dover Publications, New York, 1954.

Haldene, J. B. S., *The Biochemistry of Genetics,* George Allen and Unwin, London, 1954.

Halstead, W. C., *Brain and Intelligence,* University of Chicago Press, Chicago, 1947.

Harris, T. E., *The Theory of Branching Processes,* Prentice-Hall, Englewood Cliffs, N. J., 1963.

Hawkins, J. K., "Self-Organizing Systems—A Review and Commentary," *Proceedings of the IRE,* Vol. 49 No. 1, January, 1961, pp. 31–48.

Hebb, D. O., *The Organization of Behavior,* John Wiley and Sons, New York, 1949.

Helstrom, C. W., *The Statistical Theory of Signal Detection,* Pergamon Press, New York, 1959.

Herdan, G., *Language as Choice and Chance,* P. Noordhoff, Groningen, Holland, 1956.

Howard, R. A., *Dynamic Programming and Markov Processes,* M.I.T. Press, Cambridge, Mass., 1960.

Hsu, J. C., and W. E. Meserve, "Decision-Making in Adaptive Control Systems," *IRE Transactions on Automatic Control,* Vol. AC-7 No. 1, January, 1962, pp. 24–32.

Huxley, J., *Evolution in Action,* Harper and Row, New York, 1953.

Jeffress, L. A., *Cerebral Mechanisms in Behavior,* John Wiley and Sons, New York, 1951.

Jevons, W. S., *The Principles of Science: a Treatise on Logic and Scientific Method,* originally published 1874, reprinted, Dover Publications, New York, 1958.

Kalman, R. E., and R. S. Bucy, "New Results in Linear Filtering and Prediction Theory," *Journal of Basic Engineering,* March 1961, pp. 95–108.

Kalman, R. E., "On the General Theory of Control Systems, *Proceedings of the First International Congress on Automatic Control,* held in Moscow, USSR, 1960, pp. 481–492.

Kalman, R. E., "A New Approach to Linear Filtering and Prediction Problems," *Journal of Basic Engineering,* March 1960, pp. 35–45.

Kemeny, J. G., J. L. Snell, *Mathematical Models in the Social Sciences,* Ginn and Company, Boston, 1962.

Kipiniak, W., *Dynamic Optimization and Control,* M.I.T. Press, Cambridge, Mass., 1961.

Kleene, S. C., *Introduction to Metamathematics,* D. Van Nostrand Co., Princeton, N. J., 1952.

Kneebone, G. T., *Mathematical Logic and the Foundations of Mathematics,* D. Von Nostrand Co., London, 1963.

Kochen, M., and E. H. Galanter, "The Acquisition and Utilization of Information in Problem Solving and Thinking," *Information and Control,* Vol. 1, No. 3; September, 1958, pp. 267–288.

Korner, S., *Conceptual Thinking: A Logical Inquiry,* Dover Publications, New York, 1959.

Körner, S. (editor), *Observation and Interpretation,* Butterworth's Scientific Publications, London, 1957.

Kyburg, H. E., Jr., and E. Nagel (editors), *Induction: Some Current Issues,* Wesleyan University Press, Middletown, Conn., 1963.

Laning, J. H. Jr., and R. H. Battin, *Random Processes in Automatic Control,* McGraw-Hill, New York, 1956.

Leblanc, H., *Statistical and Inductive Probabilities,* Prentice-Hall, Englewood Cliffs, N. J., 1962.

Licklider, J. C. R., "Man-Computer Symbiosis," *IRE Transactions on Human Factors in Electronics,* Vol. HFE-1, No. 1; March, 1960, pp. 4–10.

Luce, R. D., and H. Raiffa, *Games and Decisions, Introduction and Critical Survey,* John Wiley and Sons, New York, 1957.

Lyman, J., and L. J. Fogel, Chapter 1 of *Handbook of Automation, Computation and Control,* Volume 3, E. M. Grabbe, S. Ramo, and D. E. Wooldridge (editors), John Wiley and Sons, New York, 1961.

Machol, R. E., and P. Gray (editors), *Recent Developments in Information and Decision Processes,* The MacMillan Co., New York, 1962.

MacKay, D. M., "On Comparing the Brain with Machines," *American Scientist,* Vol. 42 (1954), pp. 261–268.

Magoun, H. W., *The Waking Brain,* Charles C Thomas, Springfield, Ill., 1958.

McCulloch, W. S., and W. Pitts, *A Logical Calculus of the Ideas Imminent in Nervous Activity,* from the *Bulletin of Mathematical Biophysics,* Vol. 5, No. 4, pp. 115–134. University of Chicago Press, Chicago, December, 1943.

McCulloch, W. S., and W. H. Pitts, "How We Know Universals. The Perception of Auditory and Visual Forms," *Bulletin of Mathematical Biophysics,* Vol. 9 (June, 1947), pp. 127–147.

McCulloch, W. S., "Models of Functional Organization of the Cerebral Cortex," *Federation Proceedings of the American Society of Experimental Biology,* Vol. 6.

Miller, G. A., E. G. Galanter, and K. H. Pribam, *Plans and the Structure of Behavior,* Holt, Rinehart and Winston, New York, 1960.

Minsky, M., "Steps toward Artificial Intelligence," *Proceedings of the IRE,* Vol. 49, No. 1 (January, 1961), pp. 8–30.

Moore, E. F. (editor), *Sequential Machine; Selected Papers,* Addison-Wesley Publishing Co., Reading, Mass., 1964.

Moran, P. A. P., *The Statistical Processes of Evolutionary Theory,* Clarendon Press, Oxford, England, 1962.

Muses, C. A. (editor), *Aspects of the Theory of Artificial Intelligence,* Plenum Press, New York, 1962.

Nayel, E., P. Suppes, and A. Tarski (editors), *Logic, Methodology and Philosophy of Science,* Stanford University Press, Stanford, Calif., 1962.

Newell, A., J. C. Shaw, and H. A. Simon, "A Variety of Intelligent Learning in a General Problem Solver," *The Rand Corporation Report No. P-1742* (July 6, 1959).

Newell, A., and H. A. Simon, "The Logic Theory Machine," *IRE Transactions on Information Theory,* Vol. IT-2, No. 3 (September, 1956), pp. 61–79.

Oparin, A. I., *Origin of Life,* translated by S. Morgulis, Dover Publications, New York, second ed., 1953.

Oparin, A. I., A. G. Pasynskii, A. E. Braunshtein, T. E. Pavlovskaya, F. Clark, and R. L. M. Synge (editors), *Proceedings of the First International Symposium on the Origin of Life on the Earth,* Pergamon Press, New York, 1959.

Pearson, M. A., *The Grammar of Science,* original publication 1892, reprinted Meridian Books, The World Publishing Co., New York, 1957.

Pearson, K., *The Grammar of Science* Meridian Books, The World Publishing Co., New York, 1957.

Poincaré, H., *Science and Hypothesis,* translated 1905, Dover Publications, New York, 1952.

Polya, G., *Mathematics and Plausible Reasoning,* Vols. I and II, Princeton University Press, Princeton, N. J., 1954.

Popper, K. R., *The Logic of Scientific Discovery,* Scientific Editions, Inc., New York, 1961. Originally published by Hutchinson & Co., London, 1959.

Quastler, H. (ed.), *Information Theory in Psychology,* Free Press of Glencoe, New York, 1956.

Quastler, J., "The Complexity of Biological Computers," *Transactions of the IRE on Electronic Computers,* Vol. EC-6 (September, 1957), pp. 192–194.

Raiffa, H., and R. Schlaifer, *Applied Statistical Decision Theory,* Division of Research, Graduate School of Business Administration, Harvard University, Boston, 1961.

Rapaport, A., *Fights, Games, and Debates,* The University of Michigan Press, Ann Arbor, Mich., 1960.

Rashevsky, N., *Mathematical Biology of Social Behavior,* University of Chicago Press, Chicago, 1951.

Rashevsky, N., *Mathematical Biophysics,* University of Chicago Press, Chicago, 1948.

Reichenback H., *Experience and Prediction,* University of Chicago Press, Chicago, 1938.

Reichenback, H., *The Direction of Time,* University of California Press, Berkeley, Calif., 1956.

Restle, F., *Psychology of Judgment and Choice,* John Wiley and Sons, New York, 1961.

Richardson, L. F., *Statistics of Deadly Quarrels,* edited by Q. Wright and C. C. Lienau, The Boxwood Press Pittsburgh, and Quadrangle Books, Chicago, 1960.

Rosenblueth, A., N. Wiener, and J. Bigelow, "Behavior, Purpose, and Teleology," *Philosophy of Science,* Vol. 10 (1943), pp. 18–24.

Russell, B., *Human Knowledge: Its Scope and Limits,* Simon and Schuster, New York, 1948.

Russell, B., *Introduction to Mathematical Philosophy,* first published in 1919, George Allen and Unwin, London, 1953.

Samuel, A. L., "Some Studies in Machine Learning Using the Game of Checkers," *IBM Journal of Research and Development,* Vol. 3 (1959), 210–229.

Savage, L. J., *The Foundations of Statistics,* John Wiley and Sons, New York, 1954.

Sass, M. A., and W. D. Wilkinson (editors), *Computer Augmentation of Human Reasoning,* Spartan Books, Washington, D. C., 1965.

Sayre, K. M., and F. J. Crosson (editors), *The Modeling of Mind,* University of Notre Dame Press, Notre Dame, Ind., 1963.

Schrödinger, E., *Science and Humanism,* Cambridge University Press, London, 1951.

Schrödinger, E., *Science Theory and Man,* Dover Publications, New York, 1957.

Sebsestyen, G. S., *Decision-Making Processes in Pattern Recognition,* The Macmillan Co., New York, 1962.

Siegel, S., and L. E. Fouraker, *Bargaining and Group Decision-Making,* McGraw-Hill, New York, 1960.

Simon, H. A., *Models of Man: Social and Rational,* John Wiley and Sons, New York, 1957.

Swets, J. A. (editor), *Signal Detection and Recognition by Human Observers,* John Wiley and Sons, New York, 1964.

Solomonoff, R. J., "An Inductive Inference Machine," *IRE Convention Record,* Vol. 5, Part 2, pp. 56–62, March, 1957.

Stromer, P. R., "Adaptive or Self-Optimizing Control Systems—A Bibliography," *IRE Transactions on Automatic Control,* Vol. AC-4, No. 1 (May 1959), pp. 65–68.

Tanner, W. P., Jr., "Theory of Recognition," *Journal of the Acoustical Society of America,* Vol. 28, No. 5 (September, 1956), pp. 882–888.

Tax, S. (editor), *The Evolution of Life,* Vols. 1, 2, and 3, University of Chicago Press, Chicago, 1960.

Thrall, R. M., C. H. Coombs, and R. L. Davis (editors), *Decision Processes,* John Wiley and Sons, New York, 1954.

Tou, J. T., and R. H. Wilcox (editors), *Computer and Information Sciences,* Spartan Books, Washington, D. C., 1964.

Uspensky, J. V., *Introduction to Mathematical Probability,* McGraw-Hill, New York, 1937.

Uttley, A. M., "Information, Machines and Brains," *TRE Transactions of the Professional Group on Information Theory,* Vol. PGIT-1 (February, 1953), pp. 143–149.

von Foerster, H., ed., *Cybernetics, Circular Causal, and Feedback Mechanisms in Biological and Social Systems,* Transactions of the Sixth Conference held March 24–25, 1949. Josiah Macy, Jr., Foundation, New York, 1950.

von Neumann, J., *The Computer and the Brain,* Yale University Press, New Haven, Conn., 1958.

von Neumann, J., "The General and Logical Theory of Automata," from *The World of Mathematics,* J. R. Newman, editor, Vol. 4, pp. 2070–2098, Simon and Schuster, New York, 1956.

von Neumann, J., and O. Morgenstern, *Theory of Games and Economic Behavior,* Princeton University Press, Princeton, N. J., 1944.

von Wright, G. H., *The Logical Problem of Induction,* Basil Blackwell, Oxford, England, 1957.

von Wright, G. H., *A Treatise on Induction and Probability,* Littlefield, Adams and Co., N. J., 1960.

Waddington, D. H., *The Ethical Animal,* Atheneum Publishers, New York, 1961.

Wald, A., *Statistical Decision Functions,* John Wiley and Sons, New York, 1950.

Wald, A., *On the Principles of Statistical Inference,* Notre Dame Mathematical Lectures, No. 1. Notre Dame, Indiana: Notre Dame University, 1952.

Wellner, D. (editor), *Decisions, Values and Groups,* Pergamon Press, New York, 1960.

Whitehead, A. N., *The Concept of Nature,* The University of Michigan Press, Ann Arbor, Mich., 1957.

Whitrow, G. J., *The Natural Philosophy of Time,* Thomas Nelson and Sons, London and Edinburgh, 1961.

Whittle, P., *Prediction and Regulation by Linear Least-Square Methods,* D. Van Nostrand Co., Princeton, N. J., 1963.

Wiener, N., *Cybernetics,* John Wiley and Sons, New York, 1948.

Wiener, N., *Extrapolation, Interpolation, and Smoothing of Stationary Time Series,* M.I.T. Press, Cambridge, Mass., 1949.

Woodger, J. H., "Studies in the Foundations of Genetics," from *The Axiomatic Method,* ed. by L. Henkin, P. Suppes, and A. Tarski, North-Holland Publishing Co., Amsterdam, 1959, pp. 408–428.

Young, J. Z., *Doubt and Certainty in Science,* Clarendon Press, Oxford, England, 1950.

Yovits, M. C., G. T. Jacobi, and G. D. Goldstein (editors), *Self-Organizing Systems 1962,* Spartan Books, Washington, D. C., 1962.

Index